COOLING DOWN
PLANET EARTH

Rodney Earl Andrews

ISBN: 1482607131
ISBN 13: 9781482607130
Library of Congress Control Number: TXu 1-919-949

"*Cooling Down Planet Earth* turns Global Warming orthodoxy upside-down.

An insightful and revolutionary exploration of humankind's impact on our vanishing forests, what this has meant for the Earth's climate, and what we should do about it.

Rod Andrews' *Cooling Down Planet Earth* is a thought-provoking paean to the trees (and to the art of teaching) and a reminder that science should never be "settled" as long as there is more to learn and to discover.

Jack Wilson is the kind of teacher we all wish we had.

One comment does not do justice to this thought provoking, informative, must read book."

Bruce Emerson

"Many scientists are trying to figure out why our planet is heating up. Not Jack Wilson. He looked the other way. Why can't our planet cool itself down? Jack, in a very simple way, explains how we are preventing the planet from cooling down."

Dr. Basil Tucker

"Finally someone has figured out why the oceans are overflowing. The character, Jack Wilson, not only explains where the extra ocean water has come from but he explains how to put it back."

Josh Drain

"I picked up the book and could not put it down. I will be recommending to all my clients not to go on cruises. I will have to come up with other trips and vacations. Working vacations to help rebuild our past errors will be the next growth market.

Pierce Butler

"Jack Wilson is at his best in front of the classroom on the blackboard showing us all how we are preventing our planet from cooling. He has a workable plan. Have a seat, the front row is taken. You can squeeze in at the back."

Zackary Ryan

"I highly recommend that you read this book. It has changed forever how I look at our spaceship, called Planet Earth. I am asking all my friends to read this book so we can work together to make our planet healthy again."

Susan Sexsmith

"This book should have a warning printed across the cover. Rodney Earl Andrews is not a scientist, geographer, or environmentalist. His ability to get the messages across will have you questioning the experts. Why would anyone question the experts? Be forewarned and guarded before you open the cover as you won't be able to put it down until you have finished the last chapter. Please read this book. It will make you think and it will call you to action."

Willy Winslow

"Jack Wilson, one of the main characters, stroked my intelligence, my common sense, and my faith in my fellow humans. Be prepared, when you read this book you will be swept away. My heart, like my planet is broken. Top of my list again."

Heather Parker, Editor and Critic of
"Heather Parker Reads and Tells"
P.S. "Yes, we together, can fix our home, planet Earth."

"I wanted more and I got more, **This is a real wakeup call!!!** Thank you, thank you, **the second book** is better than the first. <u>Wood</u> there happen to be a third on its way?"

(Number one and number two on *Rudy Sweyder's List*)

"We are definitely looking at our forests differently these days. It is time we all took responsibility for our actions. Cut us in for planting our share. I am retiring my lawn mower and planting maples and oaks. We all need a bit more shade."

Tavis Tate, President *Lakehurst Loggers Syndicate*

Lowland citizens of the world are uniting to pressure the world to start planting, NOW!!! Stop cutting, you are flooding us out of our homes and livelihood."

The World Wide Confederation of Sinking Islands (WWCSI)
Future Division of Alliance of Small Island States (AOSIS)

Acknowledgements

A Big Thank You to:

Claire Archambault, my wife, who has listened patiently for the last several years during the research and writing stage of this book.

My editor Veikko Pipponen, Principal, Manxx Communications and Research, tried to make the script you are about to read legible. Veikko was kind and made lots of suggestions and comments. When you write you forget that you have readers that have to follow your thoughts and paths. Spelling, grammar, and gapping concepts are caught by your editor and you go back and make the necessary changes to keep your audience with you. When telling a story to an audience you have the visual and auditory clues to let you know you are on track. Writing a book you count on the editor's comments. Thank You Veikko.

The cover was designed by Glen Sleeman. Once again taking a concept to fruition is so rewarding when you are working with a professional artist.

The final edit was another wake up call. Once again, thank you Linda Butler.

Names, places, characters, are the product of the author's imagination and experience. Place names are used in a fictitious manner.

About the Author

I enjoy telling stories.
I hope to be judged on keeping you
awake and entertained.

This is my second book. It is so rewarding to hear and see
new friends and old friends laugh and shed a tear.
First book: *Ten Bridges Seven Churches No Stop Light*

Table of Contents

Note: You may want to read the Epilogue Interview first as it will frame your thoughts on how my second book came into existence.

CHAPTER ONE

Jack Wilson Meets John Henry

I f you had the opportunity to read the first book you've met Jack. He is the ice man. Jack grew up in a small town called Norwood, Ontario and after his education in the classroom was over his other education began. Jack paints ice in arenas, curling rinks, and outdoor winter parks. The company he owns is Painted Ice Inc. He specializes in painting on ice. If you want lines for hockey, circles for curling, or promotion graphics, he is the person to contact.

Back at the small plant and head office in Richmond Hill, Ontario we listen in on the conversation.

One morning Jack was routinely inspecting and testing the hardness of the ice in one of his many refrigerated ice trays. Sara was running the office, answering all the routine calls and working away on the bi-weekly payroll. In the shop Deloris needed help to weld two aluminum pieces.

Deloris. "Jack I notice sometimes when you're not writing in your log book you're just sitting and thinking. What are you working on?"

Jack, "Well, not having been to university and not being formally educated I sometimes get myself in trouble with my wild ideas. I guess I should say not having been to university. You know what I mean."

Jack briefly explained his Ice Box Theory on global warming and the theory he was presently working on.

As the two were setting up and clamping the aluminum to weld, Deloris mentioned that one of her colleagues in Kanata was a university professor in Ottawa. "Jack, my friend is part time and he can't get a full time position, let alone tenure, as his thinking is not in line with the head of the science department. You may not believe it but his name is John Henry. He's suffered a lot of jokes. John is a free thinker; he would be fun to talk to and a very good listener. He has spoiled me many times by just listening as I worked my way through. Would you be interested in meeting John? I could arrange it for a time when you are down Ottawa way. Be prepared, the man loves to eat."

Deloris contacted her friend and set up a meeting for a week later when Jack was traveling to Arnprior to paint a special design for the skating club.

Jack drove to Ottawa first as his time on the ice would be much later that night when the arena would be closed to the public. Jack met John at a coffee shop just off the university campus and settled in for a conversation. The staff knew John very well. Professor John Henry was one of the better tippers. A seat always could be found for John. The staff looked after the professor.

When people meet they sometimes know instantly they've connected. Maybe it's the eye contact or the tilt of the shoulders. The two men were thirty years apart in age, but Jack knew he had met another free thinker.

John: "Deloris mentioned that you had some questions and wanted to bounce around some ideas that were a bit out there. Deloris also told me how you two met when she ran out of gas and how she got the job with your company. I think she's found a home. You'll be amazed at her ability and dedication. I miss her and wish she worked closer to Ottawa."

Jack: "We're pleased to have her. She's a big part of the future. Our industry will be going through changes; we hope to create those changes before the competition gets a hold on our market."

John: "Deloris mentioned that your company paints ice. What are your future plans?"

Jack passed John a business card and a ball cap. **Rink Rats** was embroidered on the front of the cap and on the back strap it said **"Made in Canada, Eh!"** We presently paint and decorate ice in arenas, and curling rinks. We custom design graphics for bonspiels, shows, celebrations, playoffs-you name it. We paint and decorate large sheets of ice. The growing side of our business is supplying water for ice making. I'm sure you have taken old ice cubes out of the freezer and they smelled stale. We scent our water. You can have flowers, citron, or any fruit or fragrance you want. We also put in other products to change the hardness of ice. Hockey and figure skaters need different types of ice to perform at the top of their game and we provide the correct water to get the required texture and firmness. Our business takes us all over Canada and the United States and we also have a number of big contracts in Europe. We want to keep this market and also investigate other opportunities."

Jack, "I guess I am a bit embarrassed, but I observe what is going on around me, and in our business I get to travel and talk to many people from all walks of life. Some things are just not adding up and I need to sort them out."

John, "Jack, just shoot from the hip. I'll stop you if I don't understand and I'll also make a few notes so I can come back and ask questions for clarification. Is that okay with you?" John opened up his large note pad and wrote:

JACK WILSON SEPT. 14TH.

John started to eat what would be the first of six donuts. Jack was happy to pick up the bill as he finally had the ear of a professor.

Jack, "I love to talk, but I'll try to be as brief as possible. The Icebox Theory is very simple. The polar ice caps and glaciers around the world are melting because of the changed rays of the sun."

Jack had his large pad of graph paper in front of him and he started to sketch. The graph paper was cut to represent a large arena. John would soon see someone sketch as quickly as he talked.

"Air temperature increases of one or two degrees have little to no effect on the melting. Think about it John. On how many days, for how many hours is the Arctic air above freezing? If the temperature goes up one degree it would have little to no effect on the amount of ice that would melt. When it is minus 40 and it warms up to minus 39 it is not going to melt any ice. Notice all the pictures and movies you see of ice walls crumbling and falling into the ocean as the cruise ships pass by. The sun is out and cooking the ice. The on board crew and passengers are in winter coats and hats and it's cold. The beating sun, not air temperature, is the culprit."

John noted:

SUN-RAYS NOT HEAT IS MELTING THE POLAR ICE.

Jack, "Once the ice is reduced in surface area the prevailing winds are less cool and the result is a warming of the rest of the planet. A big block of ice cools more air than a small block and that is why they used to have two sizes of iceboxes."

John wondered how many students would know what an ice box was, let alone that there were different sizes.

Jack continued, "The sun cooking the ice like a micro-wave oven and reducing its mass is contributing to the world heating up. We have to block the sun melting the glaciers and Polar ice caps and I have a product that will do it. I have tried it on snow banks along a road, on the side of a ski hill, and on ice roads in the far north." Jack drew an ice road over a frozen lake in the north boreal forest and the ice on either side of the ice bridge was wide, white, and holding as a trac-tor trailer was moving across the surface.

Two students walked by, glanced at the sketch and took a seat by the window. John Henry gave a nod of recognition, to them, and smiled.

Jack continued: "The sun's rays were blocked and the ice stayed longer before it eventually melted away. We've cut down our forests which used to hold the snow and ice for weeks after open farm fields had lost their snow, ice, and ground frost. We've stripped the planet of over 75 per cent of its forest cover. If you put the thermometer in the woods you would see that the planet is actually a few degrees cooler than it was a few hundred years ago. The truth is the planet, if left alone, is actually cooling. We are interfering with Mother Nature. We have taken away the shade and we have to put the trees, the shade back.

John, "Let me get this straight. Do you think we are in a period of global warming or not?"

Jack, "You are right, I don't want to be misunderstood. There is no question we are in global warming. There are signs everywhere, the weather changes and extremes are the canaries in the mine.

Many people are looking at carbon dioxide which is good. I am looking at oxygen. We strip our planet of trees and plant soybeans that take 45 to 65 days from planting to harvest time. Trees especially fast growing evergreen trees produce more oxygen than any annual food crop we plant. We are systematically reducing the amount of oxygen our planet can produce. Look at farm fields when you get a chance and see how many days a year something green is growing. Record how many days, weeks, months the fields are brown when all around the trees are green and producing oxygen. Roads, parking lots and roof tops do not produce oxygen.

John noted: **Oxygen look at it as well as Carbon**

Jack further adds: "John, the other thing is we did not cut down our original forests around the world. The original forests have come and gone for millions of years. What we did in North America was push the first inhabitants to one side and cut the forests for the first time as we had the tools and desire to do so. Europeans and Asians in their countries had their first cut centuries before as they needed the open space to plant crops. We should have left our tallest trees but we cut everything. It will take at least four generations of trees to get back to the same height and volume of trees we started with. We have to put them back."

John wrote on his pad:

World's oxygen level, monitoring?

Global cooling, ice age, thermometers in the wrong spot, 75 % of the trees are missing. Re-measure put all the thermometers in the woods. Not original forest but call it first cut. Four tree generations to grow them back.

Jack continued, "Permafrost is melting in the Arctic. Frost comes out of the ground when it rains. Everyone that lives in the country or drives a truck and is waiting for the half loads to come off knows that. Secondary roads are all closed to heavy traffic in the spring until the frost comes out of the road bed. A few degrees warmer is not going to melt feet of permafrost. Listen, if they check out the change in annual rain they will find the answer. Rain is nature's tool to make soft land hard and to take out the ground frost."

John remembered seeing the half load signs as a small child and wondering what they meant.

More students started to fill up the seats around them and the noise level was starting to increase.

Jack continued, "Man has dammed just about every river in the world to produce electricity. Electricity in turn produces heat. Electricity is just the method we use to move power from one spot to another. This invention allowed us to move the power away from the banks of rivers."

Jack sketched a dam with cool water turning a turbine and the electrical current was flowing hot out through the lines.

John noted:

HYDRO PRODUCES HEAT FROM COOL FALLING WATER

The customers at the two tables next to Jack and John glanced over at Jack drawing on his graph paper.

"We are also burning 4.5 billion tons of coal a year and 35.9 billion barrels of oil, plus billions of cubic meters of gas. That is no small campfire. We burn millions of tons of wood to heat and cook on. We have paved millions of hectares or acres of our earth surface with black asphalt and millions of acres of roof tops that absorb heat. We have cleared millions of hectares of forest for farming. We are causing our planet to heat up because we are not allowing the planet to cool down naturally. Air pollution is a major concern. We all enjoy clean air. Air pollution gets the attention of the public. We definitely need clean air for many reasons, especially health reasons. Global warming is a direct result of the sun's rays changing and the second and very important issue is the deforestation of the world."

John wrote on his pad:

Total heat from all the fossil fuels we are burning + total heat from electricity. Planet cannot cool down naturally

Jack, "Since we have given our astronauts a decent camera-and by the way, many of those cameras are made in good old Canada, Eh!-look at the shots of earth on the shaded side. Some people call it night time. I call it shade time. It's lit up like a Christmas tree. There are very few dark spots in the world today. Earth shades itself naturally to keep cool and we light it up and heat it up. Have you ever touched a hot light bulb? Add up all the electricity we generate, all the fossil fuels we burn each day. Don't forget hydro is water powered electricity not to be confused or added in twice."

John, "Jack, I am not sure what you mean by adding it in twice."

Jack, "If you burn coal, oil, or gas to produce electricity you get two heat sources. The first heat comes from the burning process. The resulting steam is the power to turn turbines. The second heat is the electricity itself when it is transmitted and is consumed. Hydro has cool water pulled down by gravity thus turning generators. The heat is the transmission of electricity and the consumption end." Nuclear, solar, and thermal electricity, you get the idea? A bit like cutting wood. You work up a sweat cutting it, splitting it, piling it, and then when you carry it into the house." Jack cracked one of those farm boy smiles.

John nodded and smiled.

Jack, "Don't you love it when you see on your Hydro One bill a charge for delivering their product to you?" My friends call it, "Hydro One, Public Zero," or, "Hydro Won.

Jack, "You will notice that cities are warmer, sometimes by two or three degrees, than the surrounding countryside. Burning electricity is part of the reason.

The first cold day in July or warm day in January and the doubting Thomases jump on the band wagon and make fun of global warming.

Our tree cutting rate is faster than our present restocking efforts. Trees around the world help keep the earth cool. An oak or maple tree of average size, in the summer growing season, consumes over 50 gallons of water a day. How much heat is needed to evaporate that water? Evaporation of water sucks the heat up and there are only three air conditioners in the world. One is evaporation. Surface evaporation occurs wherever you find moisture. Oceans, seas, lakes, rivers like the Ouse River, ponds and wet lands are the most obvious evaporation surfaces. The second largest air conditioner is the forests around the world. Seventy five percent of the land surface was once covered in forest. Do you have any idea how much water is evaporated each day from the largest plants on the planet? Shade and evaporation keep us cool and we are continuing to cut trees down and wondering why the world is heating up. Rising water moisture forms clouds. Water vapour cools due to the altitude and condenses into clouds. It`s called the loss of evaporation. We can't change the tilt and orbit of the planet but we can put back the shade. SEA is how the planet cools. Shade, Evaporation, and Altitude are the only three ways the planet can keep cool. Seven billion humans have cut the shade trees. We are preventing the planet from cooling down naturally. Why don't people see what we are doing wrong?"

John noted:

Calculate how much water a deciduous and a coniferous tree evaporates? SEA Shade, Evaporation and Altitude are the only three ways the planet cools down! (Only three????)

John, "Jack, I have never really looked at climate change from that perspective. If humans were not on planet earth, would the planet be healthier? Investigating how the planet cools itself would be an interesting topic for a young student I have who wants to do a doctorate. Her undergraduate and master's degree would fit perfectly with that topic."

Jack, "Would I be able to read her work when she is finished?"

John, "If Montana decides to go that direction then you will be spending some time with her being interviewed."

Jack thought to himself. This is too good to be true. Imagine having someone else helping out and researching on a topic you are so interested in.

Jack, "The last time I was in Las Vegas they had fine water spray nozzles around the pool perimeter spraying water. The fine mist evaporates and cools the air so people can sit and get a tan without getting too hot."

GREEN VEGETATION WATER SINKS (GVWS)

"Trees are full of water. Ash is one tree species that has a low percentage of water and it still contains 45 percent water in the stems or trunks. John, this percent is by weight not volume. If you have 10 lbs or 10 kilograms of wood, 4.5 lbs or 4.5 kilograms is water. Leaves and small branches have a much higher water content. Sugar maple for example is over 70 percent water. Red oak is also around 70 percent water. Basswood, birch, and sycamore are up to 60 percent water. The tree branches, small limbs, and leaves even have a higher percentage of water than the trunk wood. The root structure of a tree is sometimes one third the size of the tree above ground and the roots are over 80 percent water. In a forest the ground vegetation mat is moister than the average field used for agriculture."

"The weight of a logging truck and pup loaded with logs heading to the sawmill is over 50 percent water. You've seen pictures of trees falling on a car?"

John, "What is a pup?"

Jack, "Picture a school bus pulling a trailer behind. Now picture a logging truck with the boom on the back of the truck and the trailer behind. The trailer is called a pup. The configuration of the truck and trailer allows an experienced operator to get into some pretty tight places to load and unload. The boom at the back of the truck allows the operator to load both platforms. The standard truck and pup are used when hauling veneer logs to Longlac. A tractor trailer and pup would be a disaster in the woods but are used effectively dumping gasoline and diesel at stations where you can pull through and not have to back up. Hey that may help you with a trivial pursuit question.

Think of the forest floor as a thick straw mattress. It absorbs and holds a lot of water. The ground water table is also much higher in a forested area compared to parts of the province where you see fields growing crops, one after the other.

In the winter the ground freezes in most of Canada. The polar ice cap is small in comparison to the amount of ground frozen in the northern part of the globe."

Jack, think of the forests as green water sinks. The ground water table is higher, the mattress is dense with all the plants and rotting vegetation, and there're the trees with their foliage. Forest ground cover is many times thicker than the topsoil in farm fields in Canada. All the decaying trees, branches, plants and the accumulation of needles and leaves hold moisture. With agricultural production, the six to eight inches of topsoil doesn't hold much water. John, I am talking about Canada. Most of our farm fields have a very thin layer of topsoil. The moving ice sheets of the past glaciers carried all that good soil south into the USA.

John noted:

WATER SINKS, GROUND AND ROOTS, MATTRESS, STEMS AND FOLIAGE

One large tree would have two or three tons of water. Take an acre or hectare of forest and calculate the weight of water held in the trees and extra water held in the ground. "Forests are green vegetation water sinks". Jack took out his ice marker and turned the placemat over to the blank side. He printed **GVWS Green Vegetation Water Sinks**

John also noted:

GVWS Green Vegetation Water Sinks Trees are just big vertical water tanks over 50 % full. (50 % carbon 50 % water by weight)

A young waitress soon dropped by the table with several paper placemats and nodded to Jack. "Looks like you will need some more."

John was on his third donut by now...

Normally when John was by himself in the coffee shop, former students would often sit down at his table and catch up. Today only the one couple would stop by as everyone could see he was engaged.

Sally and Fred approached the table and John said hello to Fred. Fred introduced his girlfriend Sally. Introductions were made all around and Sally continued to look at the pile of diagrams accumulating on the side of the small table.

Sally, "I attend the Ontario College of Art in Toronto and I hope to specialize in cartooning." Looking at Jack she said, "I have never seen anyone sketch as fast. Where did you go to university?"

Jack turned a bit red and felt a little embarrassed. He was surrounded by all these university educated people. Jack; "I started to sketch as a teenager and working on ice you have to be fast or you freeze. I paint ice in arenas, curling rinks, and any other large ice surface. We also specialize in large graphic designs on ice. I never thought about cartoons on ice. That would come in handy."

After a few more minutes of conversation Jack gave Sally and Fred business cards and invited them to drop in to his shop in Richmond Hill the next time they were travelling from Ottawa and were in the vicinity. Afterwards Sally and Fred took their sandwiches and cups to a vacant table in the corner. John, "Sorry for the interruption."

Jack, "I am probably the interruption." John laughed.

Jack, "Old growth pine in Ontario was a huge water sink. If you have a tree over three feet in diameter 145 feet high with its root and branch system, plus needles, it would contain over 1000 cubic feet of water. You can do the arithmetic. Canada has one third of the world's forests and we've cut them down. They're gone. My grandfather told us every time we went into the woods to cut firewood, "Do not cut your tallest trees." Self interest cut them all down and now we have wolf trees and shrubs that can't compare to the forests we exported, slashed, and burned."

John noted:

WOLF TREES?

John, "Why do you not cut the tallest trees?"

Jack, "Trees in a forest fight each other to survive. They fight for ground water, nutrients, and sunlight. Trees that have the luxury of growing in someone's front yard only competing with the grass, which is no match, spread out. The yard tree becomes a wolf tree. It grows out more than it grows tall. For example a sugar maple if left alone will be as wide as it is tall. It wolfs up the ground water, the nutrients in the soil, and wolfs up all the sunlight. If that same seedling survived in a forest it would be twice as tall and have no branches for the lower forty or fifty feet. The wolf tree is a curse, even though it looks great on a front yard. Think about it John, you could have eight or nine trees twice as high for one wolf tree. The wolf tree is good for firewood and shade, much better than a lawn. The forest tree has a trunk that can produce veneer for furniture or lumber for flooring, the list is just about endless. The top branches also produce fire wood. One large tree can lift up to 100 gallons of water out of the ground and discharge it into the air in a day. How much heat does it take to change that amount of liquid to vapour? Forests not lawns keep our planet cool."

John noted:

CALCULATE THE AMOUNT OF WATER IN EACH TREE
SPECIES AND HOW MUCH HEAT IS REQUIRED TO CONVERT
400 LITRES OF LIQUID TO VAPOUR.

"If you leave your tallest trees the new seedlings want to grow up straight to get to the top of the canopy to enjoy the sunlight. On their way up they wrestle with their neighbours and knock off each other's limbs. They try to block the sun from their tree neighbours and this causes the lower branches to be shaded and die off. Once the tree limbs die they become brittle and then the fight is on. On a windy day you can see the trees fighting each other and knocking off each other's limbs. It's rougher than football, not quite as bad as hockey." They both laughed.

"Many trees play chemical warfare with their competitors. Pine and cedar trees along with other coniferous trees put down their needles that turn the soil acidic and if they can layer enough down, many plants will not grow well among them. Other trees like walnuts have an acid in their roots to poison the competition. A healthy forest is not a park, it is a war zone among all the combatants. The strongest and the healthiest survive and grow old to die. They become a host for bacteria, worms, fungus, and food for thousands of plants.

Don't get me wrong a wolf tree is much better than no tree. People, especially homeowners, should be educated. They should be growing six or seven trees on that front and back lawn, not just one. John, you'll notice that trees growing on the outside edge of a woodlot are half wolf. On the outside they spread like a lawn ornament and you'll notice they're not as tall as the trees further in the forest. Healthy tall trees have to grow with tight neighbours and they have to struggle to get to the top."

John noted:

Grass, green lawns, are signs that we are doing something wrong.

Jack, "Thirty percent of the earth's surface is land and seventy-five percent of the land mass used to have forests or **GVWS**. We have clear cut our forests to farm, grow plants, and pasture animals. How many square miles have we put under asphalt and buildings? We're still cutting down our forests today and our global partners are cutting tropical rain forests. We beat them to it and we now have our clearcut forests growing fields of corn, wheat, hay, fruit, and soybeans. We are telling them not to cut their forests. We clearcut first, now we don't want them to follow our example. Talk about a double standard!"

"We should be providing rubber boots for the people at sea level. We cut our trees. The water ends up in the oceans and people living at sea level have flooding problems. Is that fair?"

John noted:

Historic data for forest depletion, data of ocean levels, compare??

John, "That would also be a good research topic for one of my classes."

Jack, "Cutting down our forests and spilling the water back into the oceans is just one water sink that was emptied. There are more water sinks that we have also emptied."

"In Ontario alone we share the largest fresh water lakes in the world with the United States. The volume of the water in the Great Lakes is smaller in comparison to the water we had in our original forests. We have reduced the **GVWS**. United States, China, Europe, and Russia, the largest land masses have destroyed the largest GVWS in the world. We wonder why the ocean levels have risen and are still rising."

John noted:

Canada, USA, Russia, and China calculate how much water is gone back into the oceans from each country?

Jack took a breath and sketched with a graphite pencil an outline of a tree and with a blue marker filled in half the trunk, branches, and leaves with his blue marker. "John, most trees are half full of water and we cut them down. Guess where the water ends up?

John noted:

Plant Trees, Recapture the Water, Lower the Level of Oceans

Jack, "Melting polar ice caps on land plus receding glaciers are a contributing factor. They have a great visual impact for catching the attention of the population. Polar ice caps are very visual, but, no one sees the trees, as vegetation water sinks of the world. We are all in charge of trees. We can do something about it." We need help to draw attention."

Jack was reaching for his fifth donut. Jack caught the young girl's eyes and more donuts were on their way.

Jack, "If you graph the reduction of the World's forests over time there is a direct correlation with the rising water levels in the oceans. Everyone has ignored the water content of our trees. They are huge **GVWS** and they are being emptied."

Jack, "If you look at all the countries in the world it's bad news. Holland has dumped a lot of water into the oceans and they live below sea level. Dykes hold out the water. How many acres of water did Holland alone pump into the ocean? Florida was the largest wet land in the USA. Entrepreneurs over time have cut down 98 % of the forests and drained the wet lands. Florida today is the second largest producer of beef cattle in the USA, and if climate change continues and Texas becomes the biggest desert, Florida will be the largest beef producer. Calculate the water that was stored in the huge forests and is now in the oceans. Ditches, trenches, and pipes have been put in place to drain Florida. Florida is the world's biggest example of unconscious development. To make a golf course you put in an excavator and dig out ponds and use the fill for fairways and greens. Houses in Florida in 90 % of the state have no basement as the water level is too high. Florida is a drained tropical forest swamp. Parts of Florida have had so much water drained that the land mass is now below sea level. Sink holes are caused by draining out the ground water. Sink holes are swallowing whole houses as the water is being pumped out of the ground faster than the replenishing rate. We criticise Brazil, Argentina and other countries for cutting their forests but you have to look at what we did first."

Jack, "I am starting to repeat myself. I just get so upset that we were so, I don't know if I should use the word-but "stupid" fits in this case. Ask anyone in the cold climates of the world. Frozen ice on top of water can hold a car, a truck, a tractor trailer loaded with goods heading north. Remove the water under that ice and you will have a sink hole instantly. You will notice the pictures of the sink holes that they show in magazines and on the news have no water in the holes. The cavity left when they drained out the water is the same cavity if you took the water out from under the ice. Stop taking out the water."

John, "The next water sink is the aggregate water sink. I use the short form AWS. Oak Ridges Moraine, Norwood Esker, and the list goes on. The Earth's surface holds water. Trees and vegetation on top of the topsoil hold a big percentage of the water. Topsoil holds moisture then there are the big world sponges. These sponges are deposits of aggregate, or gravel and accumulated layers of topsoil. Topsoil in some countries like Argentina and Brazil can be 80 to 100 feet thick. The top soil holds a lot of moisture. In Canada we are lucky to have 6 to 8 inches of topsoil. What we have in abundance are gravel deposits. The esker in Norwood is the gravel bottom of a river that ran under the glacier during the last ice age. Drumlins strewn around Peterborough County are another good example of mixed aggregate. These huge natural sponges are aggregate water sinks. They are being mined and turned into concrete and asphalt. Concrete is used in a thousand different ways and in every case once the concrete dries the water is off on its way to the ocean. Asphalt roads are dry layers of material. Our road systems are not only the worst examples of clear cutting they are deserts. The roads are built to not let water seep through in the summer or winter. You have dry strips running for miles-millions of miles in Canada alone. We have to change how we build roads, parking lots, and millions upon millions of asphalt driveways. We have to allow the water to penetrate and stay. We can't continue to dump it back into the ocean."

John noted:

AWS aggregate water sinks calculate the millions of acres paved over, roads parking lots, driveways Joni Mitchell is ringing in my ears

John made a mental note. It has been a long time since he heard someone Jack's age so passionate about a topic other than politics. This calculation would also be a good essay topic.

Jack, "How do we stop this and build back these huge sinks? We need a name like Dr. David Suzuki to be a spokesperson".

Jack, "Sometimes you dream of meeting someone famous and, well, I guess I'll have to let that dream go just for now. My wife tells me I never let anything go. I'm just a stubborn Irishman. Maybe someday I'll get to meet one of my heroes. Sure would like to shake his hand."

Jack, "What about Al Gore?

Trees hold a lot of carbon so we should be planting more trees for both carbon and water sequestering."

John; "We will find someone that can champion the cause."

For the first time, Jack felt that he was on the right track and someone important, besides his wife and son, was finally listening to him.

John; "Jack where did you get your information about the water sink in trees?"

Jack; "I grew up in a small town. Norwood had a population of 1000 and many were retired farmers. When our family went to the sawmill I was young and too little to shovel sawdust for the ice house. We would get at least three truck loads a year. I would hang around Billy Gallagher the sawyer. His mill was on the east side of the Ninth line just north of

Norwood. There is a trucking company, Archer, there today. Everyone else would be busy loading sawdust."

"He told me one day, Jack I don't know if I should call you Count or Thousand."

"I asked him why?"

"Billy said, "I have never met someone who has more questions or counts everything. You count the logs, you count the teeth on the saw, and you count my oil cans, my cant hooks, axes everything."

"So, Thousand doesn't sound right so you are The Count from now on."

"When I was very young I smelled and felt the sawdust from fresh logs and dried logs. Dad always said, "Shovel the dry sawdust, leave the wet or green sawdust." Logs were full of water and the sawdust and boards came out of that big saw wet. I have seen green logs squirt liquid when they were clamped in place."

I asked him, "Mr. Bill what is that stuff squirting out of the logs?"

"Sap", he yelled over the whine of the saw and the whistling of the counter weight, "Stand back, that blade eats small boys. Don't go near that belt. Hey Count, take that red marker and put an X on the end of each log and let me know how many logs Fred just dropped off."

At lunch Mr. Bill would be busy talking to himself. I just happened to be there and would ask him a few questions. I found out that the early saw mills were located on a stream or river so the water could be dammed up and used as power to turn the wheels or turbine to run the machinery. Saw, planer, conveyor belts, sawdust chute all needed power. Mr. Bill took me down to Hope Mills near Keene, close to Lang Pioneer village, and we had fun talking to Mr. Bill's friend who was the sawyer. Logs were transported to the mills by water in the old days and some logs were so full of water that they sank. Mr. Bill said that is how the name waterlogged came about. There are spots around the world where logs stopped floating and sank. The wood and the lumber sawed from these logs today is as good as the day it sank.

The outside bark and a bit inside is mush but the rest is good wood. We stopped at one of Mr. Bill's friends and had a tour of the sap house. Since that day I have been hooked on maple taffy cooled on fresh white snow. You know in the spring farmers can take 40 litres of sap out of an eight inch maple and not kill it. Sugar content is higher at the first of the run and birch sap has a lower sugar level than maple. I nearly wore Mr. Bill's ears off that day. He and his wife did not have any children and I guess he just liked me. I ended up delivering the Peterborough Examiner to them on my route when I got older. They lived in a small house on the way to Havelock."

Jack drew an old barn board shed on the side of a small but fast moving river. Logs were floating in the pond above the dam and the large side doors of the shed were open. You could see the large circular saw inside halfway through a large oak log.

Lumber was coming out one door and piled on a wagon. Slabs and edgings were pushed out another door and a conveyor was pulling out the sawdust and dumping it onto a large pile.

"In the old days the streams and rivers were needed to transport the logs to the mill but what I didn't know is that the water kept all the bugs out. Pine logs can be chewed apart in less than a year; oak a bit longer. The logs floating in the pond also stayed wet and the ends did not check. Logs that got water logged sank and some are still there today, waiting to be lifted and turned into lumber. It was also a bonus if the bark fell off. Mr. Bill said in many locations the fear of fire stopped many mills from burning their sawdust so they just put the chute into the river. Imagine today if you did that! Because you did it in the past does not give you the right to do it in the future. I read that somewhere.

Some green logs did not float well so in the boom they had to be put on the top layer. Green logs are full of water when they are cut and many sank. Everyone in the logging industry-lumber industry-knows trees are full of water. They just have not made the connection. When the water evaporates out of a log or piece of green lumber you don't see it leaving or worry about where the water went. Out of sight out of mind. Green trees and logs weigh a ton. Have one fall on your house roof or car and you'll soon understand the weight of carbon and water called trees."

John; "I've had too much coffee. I'll be running beside the truck on the way home. I may have enough energy to pick up all the timble weed on the way." Jack noticed John's eyebrow twitch. "Timble weed is the name we give to all those disposable cups and lids that blow into the ditch."

Jack; "You know it all came together when I read that if the polar ice caps melt, the oceans will rise 16 feet. The numbers didn't add up for me. First, ice floating when it melts does not increase the water level. If it did, every bar in town would have glasses of drinks overflowing. When you look at the globe and put the poles in proportion to the oceans you'll soon realize that to lift the oceans 16 feet, the poles would have to have miles of ice on the land portion. Then I said to myself, why have the oceans been increasing gradually over the last two or three thousand years? We certainly are not getting water from another planet. The water had to come from the land mass. If the poles have not been melting, where did the water come from? Since the days when humans decided to stop being hunters, gatherers, and started to be major burners, we have been on an environmental downhill slide. They have turned to farming. Settlement after settlement started cutting down trees and planting crops. Over 75% of the forests have been removed completely and the forests that are left are shrubs compared to the original growth. Only a few pockets in the world are left. Those areas that could not be logged economically are what are remaining of the tall forests.

When you work around trees in the forest and lumber industry every person that has picked up a stick of firewood or a piece of lumber knows the difference in weight between the two. Then presto, I put it all together. Where did all this water go? Drying firewood or lumber the moisture is wicked up into the air like the shower in your home and it goes downwind and forms clouds until it rains or snows. The trees are the water containers of the world and we have removed them. Out of sight out of mind. On the plains and steppes around the world that were overpopulated by wild grass-eating herds, the trees were eaten or had their bark rubbed off and killed. In Canada we were very smart. We replaced the buffalo with cattle and continued to overgraze and keep the trees off the prairies. Overgrazing, over-farming, it's the same. Humans have taken the trees off the planet. John, we have to find a way to put them back, trees are green vegetation water sinks and must be replaced."

Jack was on a roll. "In Europe and other grape growing areas of the world they cut down the forests. France cut down their oaks; Germany, Italy, Portugal, and Hungary all cut their forests to grow grapes. Countries, like Chile, and Australia as well as others did the same. Compare the water content of grapes vines to a forest of trees."

When a tree is broken and falls down or when a branch is cut or ripped off water does not leak out. The only exceptions to this are birch and sugar maple trees in the spring. People see sap processed into maple syrup. They don't think of the sap as being water even though you have to boil down 20 to 40 litres of sap to make one of syrup. Even the syrup contains water."

Jack, with the minimum of strokes, was sketching vineyards, fields of grain, and large oak trees in France falling.

Jack handed John a stack of papers all neatly labelled and stapled together. One article was on top.

"The difference between free water and bound water: even though it is self-explanatory, it is useful to think of the wood as a sponge. Once the sponge is squeezed out (free water) the sponge is still damp and needs to dry (bound water…Wood for furniture and lumber is dried down to about 6 % then after the product is in your home it will increase and decrease with the moisture management in your home. You will notice in the winter the wooden floors shrink and creak and then in the summer they swell back and the cracks are gone. Storing good hardwood furniture in an un-heated building will soon loosen the joints, dry out the frames and cause it to fall apart faster than in temperature and humidity controlled environments." S & W Report, August 1, 2011, Chris Lincoln.

Jack continued; "John these are my notes and calculations and I hope you can look over them and see if I have made any errors. A board foot is one inch by 12 inches by 12 inches so a cubic foot is 12 board feet. Firewood is 4 x 8 x 4 = 128 cubic feet. If you take all the air out between the blocks of a cord of wood there would be close to 85 cubic feet of solid wood. There would be as many branches as there would be cord wood per acre. The water content of logs is between 45 to 70% as each species has a different moisture content. Canadian Ash is one of those trees that has a low percentage of water. Do you know it's one of the few trees you can cut and burn on the same day? Good to know if you run out of wood in the winter. I got the number of acres of original forest and the number of acres today and you can see we have cleared out our forests to plant crops and to build roads and houses."

Jack; "What about that big name professor at your university. Would he be interested in my views?"

John; "I've seen many professors hired just because of their names because the university needs the recognition. They're usually a disaster in the lecture hall and they're antisocial with the faculty. They are worth every cent they are paid because the public is swayed and enrolment and funding from alumni increases. I don't think our big name professor is interested in other people's ideas. The guy is a genius. He's in a different world than you and I.

My best friend Peter nicknamed him Professor Loof Lirpa. Try reversing the name without smiling.

My colleague thinks he is like an old box of cereal; contents may settle in shipping and there is nothing at the top. All jokes aside, he's brought money and respect to our university. He is worth every cent he is paid."

Jack looked at his notes.

"John when I look at a tree I see a column of water 50% the size of the tree. John we have to build back our forest stock around the world to hold the water. Our experts are focused on carbon sinks and photosynthesis, oxygen producing plants and trees, and that is important. How do we jolt them into the importance of the green vegetation water sink?" In Ontario, like most places the biggest problem is electrical distribution. Along most roads they have their wires and we need to plant trees. Poles and wires initially was the cheapest way to go but we are paying a big price for this system. We have the equipment today to put all those hot wires underground. They're busy heating up the planet with the electricity lost in transmission, plus they'll block tree planting. Their wires are in the way of our future. Imagine how many trees could be replanted just along our highways and roads!"

John, you have heard the expression. You can't see the forest for the trees.

John noted: More than **Carbon footprints, trees store Carbon**

John was busy taking notes as fast as he could. It was hard for him not to be punching the calculator on his laptop. He was turning the numbers in his mind. How many hectares of cropland clear cut? How many hectares of forest cropped too heavily? The weight of water? The calculator was keeping up. The second large order of chilli had disappeared and muffins would soon be at the table.

Jack; "My cousins in Norwood, Evert and Rose have the same problem with their windshield wiper patent. The system we have in place today is blinding the drivers and they don't know it. Thousands of people die each year and the industry has decided making money and shareholder value is more important than lives. John, that topic we'll save for beer and wings some night."

"We are cutting down trees and raising the ocean levels. How do you get the attention of the spokespeople of the world?" Our green vegetation water sink is in front of us every day as we walk out of our homes. Look at a tree and see water in storage. We need our scientists to see this huge water sink that we're cutting down and reducing every day?"

John; "I walked right by it. This may interest a lot of people. I never thought about the green vegetation water sink and its role in holding water let alone its role in cooling down the planet. This could be a way to catch people's curiosity."

Jack checked off another bullet on his notes. "Rising ocean levels will bring legal action. Building higher dikes, flooded coastal land, a drowned Venice, and the list of world problems goes on. The economic cost and potential for loss of life will have to be rectified."

Jack, The day is fast approaching when the flooded islands and countries will be going to the Hague and laying charges against the nations that have removed their forests and the countries that are still cutting. With a smile Jack continued. The Dutch dyke and pump experts will not have far to travel to court. The court will have to determine the amount of money awarded to the countries getting flooded. The past history of buying carbon credits will seem like a minor expense. I have this picture in my mind of troops from the Caribbean and India, armed to the teeth, guarding the forests in Russia, Brazil, and Canada. Lawyers and courts on the international stage will be suing. Russia, Canada, China and the USA are the first prime class action targets.

The largest forest clear cutters are our governments, followed by electricity providers. All those roads and wide swaths cut across and through our land. What a disaster. The governments and their wholly owned subsidiaries, like Hydro One, will be sued first as they are the largest clear cutters in every country. Pipe line owners are another clear cutter.

Previous provincial governments killed the province owned and operated nursery tree business and have left it up to the marketplace to propagate new seedlings. "....complex management system was gutted resulting in a massive reduction in annual tree planting---down to fewer than two million trees per year from a high of 20 to 25 million in the 1980s.... " (Canadian Geographic, June 2010, p 55.). Governments tax the logger's stumpage and generate millions of dollars. They only invest a small percent back into forests. They do the same with fuel and the fuel tax. They only spend a small fraction of the money collected back on the roads and highways. They also subsidize farmers to irrigate and install drainage systems to further reduce the water sink." Canadians don't trust their government and for good reason."

Jack, "Again, from the Canadian Geographic magazine, page 56, June 2010: it reads, "Landowners have been trained over the years if government says to do something, they're not going to do it!""

Farmers put livestock into pastures, and the animals trample 25 % of the crop. If we want livestock we should have to feed them in forest runs. We are misusing our precious space on earth. Governments should be encouraging farmers to turn the wet fields into forests to hold more water.

John noted: **Forest runs for cattle?**

John, "Politicians know how to get elected and then they support the people and companies that helped them get the seat. The old "Golden Rule": the person that holds the gold makes the rules."

Jack, "Would you like some more coffee to wash down those muffins?"

John nodded and Jack ordered one more coffee and two bowls of chilli.

Ground Water

Jack, "Oh, how is your time? I'll be leaving here and going to Arnprior to decorate the ice for a skating carnival. The rink closes at 1:00 am and I will have the ice until 7:00 am as they cancel the first two morning hours to give me the time to paint. They need to put a few floods on before the skates hit the surface."

John, "Not to worry, D told me to take the day off."

Jack, "Some of the experts today are thinking in a trench. They see that melting ice creates water and warmer water expands. What they don't realize is that we've been pumping ground water out of the earth's surface for years. We use it and flush it back into the oceans and seas of the world. Imagine going to the moon and drilling into the surface and bringing up water, using it then letting it float away. All countries in the world are guilty of decreasing ground water and if someone took the time to measure all the ground water we have mined and flushed into the ocean, we could account for a fair share of the rising ocean levels.

In Ontario there are detailed records kept for every well drilled. The records indicate flow rate and, most importantly, the static level of the water in the well. As you pump the water out of the well, the static level temporarily changes in relationship to the recovery rate of the well and the volume of water you are removing. Measuring the static level over time you'll find that in many parts of the country the static level is dropping. When the static level drops you're removing ground water faster than the rate of replenishment."

John noted:

DRILLED WELL STATIC LEVEL CHANGES, IMPLICATIONS

PhD THESIS TOPIC FOR SURE

Jack continued, "Artesian wells are different. Water enters the aquifer above the well, sometimes many miles away. At a lower level, when you punch through the non porous stone layer, the water below that is under pressure flows out of the hole. By capping and measuring the water pressure over time you can determine if you are depleting the underground water. Just as a quick aside, on October 11, 1658 Jacques Archambault was a French Colonist and the first to drill a well in Montreal. Next time you're in old Montreal you'll find a monument-drinking fountain no less, in his honour. Very appropriate and also very useful on a hot day in Montreal.

John, there's another problem around the world and Norwood is no exception. We're pumping out ground water faster than nature is replacing it. This excess pumping of water is also helping overflow the oceans. In Norwood for example, most people in town had dug wells and outhouses just like on the farm. I grew up with a two holer. A lot of good conversations took place in the outhouse. Over sixty years ago the town put in a very large well and started to pipe water to the village homes and businesses. Toilets moved inside the house. If you've ever had to use an outside toilet in January you'll understand. Flush toilets required cesspools then septic tanks. Later, large weeping beds were required to carry away the water. Water was so cheap and plentiful that houses did not have water meters. You could use and waste as much as you wanted. Once the ground was saturated and causing problems, a town sewer system was put in. All the water today is supposedly treated and it's dumped back into the Ouse River. This is the problem. The ground water is being brought to the surface, used, then the sewers carry it away, dumping the treated water into the Ouse River. The Ouse River drains downstream to Rice Lake, and then eventually flows into Lake Ontario. The small

Ouse River, carrying this flushed toilet water from Norwood, eventually makes it to the Atlantic Ocean. Everyone should know where their waste water ends up. All the dishwashing, hand washing, and, showering soap containing phosphorus makes its way to Rice Lake. This is just one system that has destroyed all the rice beds in the lake.

In the Niagara region where you will find high concentrations of clay soil and in towns like Lakefield, Bobcageon and Kingston that are sitting on limestone shelves, the flush toilet and septic systems soon saturated the thin soil layer. All those sewer systems dump into the same watershed.

The Norwood town well is pumping water out of the esker aquifer. When the town well first went in, there were a number of small ponds a mile or two outside of the village. Now these ponds and swamps three and four miles from the center of town have dried up. The water table dropped for miles around. Many people, mainly farmers, outside of town had to abandon their dry dug wells and call in drillers to go deeper to find a supply of water. One dairy cow drinks over 25 gallons of water a day. If you want milk you have to have a large supply of fresh water. I also have to mention that most of the esker has been removed as eskers are full of valuable gravel. If you look at the esker east of Norwood you'll see the most ugly landscape ever. The large metal hydro towers have had the esker removed all around their base and they are standing atop little hills of gravel. Norwood is just one small village and how much water in sixty years has the aquifer lost? All that ground water has made its way to the ocean. Every small town and village in the country has their own story on how they cut down their trees and pumped the groundwater table down.

By the way John, there are only three professions that still get paid if they are wrong and if my wife comes back to this earth she wants to be one. Economists, weather forecasters, and well drillers, well drillers just keep on drilling and keep getting paid. I'm sure there is water down here somewhere! Economists are great at explaining to you what happened and weather forecasters look out the window and can tell you if it is sunny, raining or snowing. After that there're not very accurate. Even when they're wrong they still get paid."

John noted:

Ground water depletion overflows the oceans, how to tick off buddy Dr. Paul, the resident economist. Well drillers, forecasters economists

Jack took a bite out of his donut. "Chip, my son, came home from a job in Phoenix, Arizona and he said the whole state is pumping ground water into the ocean. First they pump fresh water out of their limited aquifer. They use this precious water for lawns, flushing toilets and the list goes on. All that water ends up back in the oceans and they continue to expand retirement communities. More people, each winter, come and consume even greater quantities of their aquifer. New land use policies are needed to reverse the flow of people especially retired people to more suitable locations. Houses and factories should only be built if they put in the proper equipment to be off the water and sewage grids."

John added to his note pad: **Move people to more suitable locations.**

Put it back

Jack continued, "We need world laws that are enforceable, that will put in place the mechanism to fine water abusers. Groundwater must be restored. Using Norwood as just one example, the extra surface runoff each year by the Ouse River would be more than enough to fill the aquifer. One problem is the aquifer is almost gone. It was sold for profit and is presently in concrete walls, floors, asphalt driveways, or highways. The Trent Severn Waterway has a problem each spring managing the dams to prevent flooding. All that excess water eventually gets dumped into the Bay of Quinte in Lake Ontario and flushes out the St. Lawrence River to the Atlantic Ocean. We need a system of injecting the spring water back into the aquifer. Common sense, which is not common, should be that if you are drawing water out of the ground you have to put it back. In our watershed alone there are hundreds of examples where the ground water is being depleted. There's not one location where they are putting the groundwater back.

What is a word we can use for putting back water into the water table? Well, is the word we use for digging or drilling and then sucking the water out of the water table. "Well" has a connotation of being well. Sucking out the water and not putting it back is not well. Maybe we should rename wells "suckers," and when we put water back into the water table we call them wells?

You can't finish if you don't start:

John the spring runoff in the Trent Severn watershed is enough to fill every bottle of water sold in one year for the entire North American continent. We just flush it every spring into Lake Ontario and out into the Atlantic. I met Douglas Barnes. He is president of EcoEdge Designs Ltd. The company designs systems to slow down the loss of water and by contouring our land and tilling it correctly you can reduce summer droughts and store much needed spring water for use later in the growing season. Douglas should be hired to change our road ditches. The thousands of miles of ditches alone if properly designed, along all the roads in the watershed, would prevent spring flooding and add much needed water back to the underground water table. Douglas is the guy they need to hire to start putting in place a solution which we will all benefit from.

Jack, "The Ontario government subsidizes farmers to drain wetlands to increase crop production. Farms with low-lying wet fields can't grow crops that need a long hot growing season. Trenching and putting in plastic drain pipes will allow the water to seep through the overlying soil into the pipes and flow away. The land will become workable early enough to give time to plant the crops that will give the farmer the greatest return. The farmer pays a certain percentage of the installation and the provincial government pays their portion. Sure enough next spring the snow melts the rain falls and the drainage systems works and the fields are dry enough and can be planted. Draining the fields allows the heavy farm equipment to prepare the soil and plant the crop. Farmers in their own self-interest think this is a good idea. Land is expensive and this wet field if drained can generate more revenue. Here is the catch. We all pay for this system. The thousands of acres that are draining due to government grants are flushing spring snow melt and rain water into the Trent Severn Waterway system and thus increasing flooding problems in the whole system. The provincial government by subsidizing draining of wetlands is creating more water for the federal government, TSW, or Parks Canada. The problem does not end there. Farmers have been educated to keep their livestock from drinking directly out of streams, cricks, and lakes. A cow will stand along the bank or leg deep in the fresh water and as they drink will defecate in the water at the same time. Putting up barriers to prevent cows from drinking directly has worked. Meanwhile you irrigate a wet field with drainage pipes."

John interrupted. "Oh shit I know where you are going with this line of thought."

Jack grabbed his note pad checked off another bullet and continued. Farmers spread liquid manure and solid manure on the fields. They then plant roundup ready genetically modified corn from Monsanto with roundup ready chemicals. The rain runoff flushes these chemicals and manure into the drain pipes. You know where these pipes end up. Shit runs downhill and always will. Downstream people are drinking this water.

John, "roundup ready?"

Jack, "Seeds have been engineered to grow plants that are not affected by a killing spray called Roundup. Roundup kills all green plant life but will not affect the engineered seed so you can see why this is a great product for the farmer but not for the environment. Farmers used to be land stewards, but sadly today they are just trying to make a living and producing food as cheaply as possible to get to the next growing season. We all pay a price for this. Farmers know they are killing the bees and destroying their soil but, well I don't need to say any more."

Get this, our drainage system was created by nature when the land was forested. Subdivision after subdivision, parking lots, roadway after roadway, take down our forests, now add thousands of acres of wet farmland that is draining into the Trent System. Get on the phone and blame Trent Severn Water Way for the flooding!!!! The federal and provincial government both own this problem.

You can look up the records and calculate the thousands of acres that have been drained thanks to the subsidy programs of the provincial government. Imagine if those fields had been turned into wetlands and in the spring they had small dykes to hold even more water. What would be the impact on the ground water table and flooding of the Trent system?"

Jack, "John, I'm really good at telling other people what to do. Every house and business in town should have to install, or in the older homes reactivate, the cisterns. With the pumps and plumbing systems we have today all the rain water could be used. Imagine having soft water to wash in. Put a big pump beside the Ouse River and run a large water line along the top of the esker. Drill a series of large holes down into the esker. Line the holes with a porous pipe and in the winter and spring put back the water. It would be easy to calculate how much water is being pumped out of the aquifer each year and it would be easy to calculate how much you are putting back into the aquifer. It would be impossible to replace the gravel that has been taken out and used. You know on the north side of town they're using one of the old gravel pits as a dump. Think about it. All the rain and melting snow filters through the garbage and into the aquifer where Norwood draws out their drinking water. Every village has its own story and their own mistakes"

John noted:

Cisterns...put back the ground water storage system. Wells are suckers?

Jack, "Think about countries like China, India, and most of Europe. They're digging deeper and deeper to find ground water. The classic story we all know are the two Chinese farmers that were digging their well deeper and they found the pottery that led to the discovery of the terracotta soldiers."

Jack did not mention to John his thoughts about watersheds. Jack thought we shouldn't be shedding our water. We should be trying to store, conserve, reuse and slow down the water leaving our drainage basin. So far everything man has done is to do the opposite.

❦

Geo Thermal is a Non Renewable Resource

After a few gulps of coffee Jack continued. "The other major problem we're faced with is the idea that geothermal heating is a renewable resource and a free source of heat and energy. We've only started to tap into the crust of the Earth.

Seven billion people around the world tapping into the crust will start to cool the centre of our planet. When it cools will we all die? Our lakes and oceans will freeze from the bottom up, and the sun will evaporate the ice as the atmosphere disappears along with the vegetation. North Americans have been able to pollute the Great Lakes in a few short years. Our oceans and seas are all in trouble and they cover over two-thirds of our planet. We can soon punch enough holes in the crust to start reducing the internal heat of our planet. Just give us a few more years and we'll screw it up.

We've polluted the oceans, Great Lakes, and most bodies of fresh water in two generations and we can cool the centre of our planet."

John, "I have to check my numbers and run them by a few colleagues discreetly. I'll get back to you within the week."

Jack, "John, can you talk to someone in the economics department?"

John, "Sure. What would you like me to ask?"

Jack handed John a chunk of paper he had cut out of a magazine being discarded at the dentist's office.

THE INVISIBLE HAND

The invisible hand is a term used by Adam Smith to describe the natural force that guides free market capitalism through competition for scarce resources. According to Adam Smith, in a free market each participant will try to maximize self-interest, and the interaction of market participants, leading to exchange of goods and services. This enables each participant to be better off than when simply producing for him/her self. He further said that in a free market, no regulation of any type would be needed to ensure that the mutually beneficial exchange of goods and services took place, since this invisible hand would guide market participants to trade in the most mutually beneficial manner.

The invisible hand beats a handout.

Jack, "John this way of thinking has driven our society to a state where the seven billion rats, I should say seven billion humans, working in their own self-interest will perish together. They are consuming the planet they are living on, fighting and killing each other and not having much fun doing it, except for the king rats."

I need someone educated to write a rebuttal, I guess that's what you would call it.

I would say it like this: It is the Visible Hand.

In a world we all live in, everyone must think and act in such a way that all people benefit from our actions. Regulations with controls are necessary as a parent is necessary to teach their children responsibility and respect for the world they live in and the neighbours they live with. The global village must raise the next generation to care, take the knowledge base we have accumulated, and to champion the individuals and groups that broaden our understanding of our future. We must work together to bring the human population-the invasive species-under control. Every person, every day must feel safe, be fed and housed and cared for, and be a part of our community."

Jack continued, "Our resources are not scarce. There is enough for everyone, and we must live by one rule only: when in doubt do what is best for everyone in our family, our neighbourhood, and our global village. We are on this journey together. At all costs we must nurture and protect the people and the planet we presently live on.

We will not need handouts; we must justly deal with holdouts. Presently we are seven billion rats floating on a piece of cheese and we are eating it.

Jack my guess is that the wheels will turn and the world will wake up that we are in trouble. Laws will be put in place to stop burning gas, coal, oil, and wood products. The negative economic impact on OPEC and the new positive for the solar industry will be mind-boggling. Will we survive the economic crash?"

John, "We will survive if the king rats can stay on top of the new pile. They will protect their self-interests first and foremost. They will change the world order only if they can gain from it. Lost my tenure again."

Jack, "I think we have to start now. My dad always said you can't finish if you don't start. He watched too much baseball."

John, "Jack do you have a theory on how the Earth was born?"

Jack, "Yeah, but that's for a big pitcher of beer and a pile of wings."

John, "You're sure about global warming?"

Jack, "Cut down eighty-five per cent of the world's natural air conditioners depleting the world's largest green reservoirs of clean water, pave millions of miles with black asphalt, build millions of homes and factories with dark roofs, burn four and a half billion tons of coal a year, burn over ninety-eight million barrels of oil a day, burn millions of cubic feet of gas, burn millions of tons of wood, dam up every moving body of water on the planet to generate electricity, mine uranium and produce more heat, tap into the earth's crust and generate heat, harness the wind and sun-even more heat. The seven billion rats are eating up the planet but at least they will be warm until the end. Then it will be really warm.

We are the invasive species, out of control, with just enough brain power to destroy our space ship Earth before we understand we are on it together.

Global warming, you bet. Move close to the campfire and we're all roasting."

Jack just smiled. "We are all in this together. You know I'm wondering if we have enough grade six students to explain to the adult population what's happening."

John, "Jack, is there any chance you could stay overnight and we could talk tomorrow. My mind and lap top are full and I need a break."

Jack, "John, you're as thin as a rake and I can't believe how much you can eat."

John, "When I was a teenager I was so thin I had to stand up twice to make a shadow. Got married, gained a few pounds, so I'm working on a new theory: If you eat a lot you lose weight carrying it around. It works for me."

Jack, "I'm back next week to paint a curling rink in Pakenham and we can get together - I hope. Lets meet at a local coffee shop."

Next week in Pakenham.

John met Jack at the agreed upon local coffee shop.

"Hi Jack; let me buy the coffee and sandwiches today."

"No John, I get to expense this kind of stuff and you don't. Besides, have you heard the joke about teachers?

What is the difference between a teacher and a canoe? One tips!"

They both laughed as they knew the stereotype fit the profession well. When Jack worked as a teenager at a summer lodge the owner booked staff parties in the spring and early summer. If it was a group of teachers he would always add fifteen percent to the price, as they never left a tip for the staff, unlike the other company groups. Lawyers say that judges have deep pockets but short arms.

John, "Jack, I reviewed my notes and read them over a number of times this past week and I've had time for your ideas to sink in. I believe you've solved the mystery of why the oceans are rising and you have also solved the reason for the planet warming up. Solving these two problems are the discoveries of the century. Houston we have a problem!" They both laughed.

Jack was very excited and could not quite believe his ears.

"John, why does Houston have a problem?"

John explained, "Your solution is too simple and I anticipate the scientific community will have trouble believing it.

Think about it for a moment. Thousands of university professors around the world and thousands of scientists have spent countless years and billions of grant dollars to unlock the mysteries of global warming and rising ocean levels. We have cut down forests to get enough paper to supply the magazines, newspapers, and photocopiers for the countless articles that keep focusing on the carbon footprints. Being blunt Jack, you do not have the credentials nor do I."

Jack: "What's wrong with a good idea and common sense?

John: "Your ideas will rock the cash boat. They receive grants, studies, donations, sponsorships; for some the list of perks goes on. You're not only turning the boat around, you figured out why it is sinking, found the leak, and now you must convince those upset passengers to help you plug the hole.

We have to have a plan. Somehow we have to break into the scientific circle, gain acceptance and then let the scientists know what you have discovered. It may take some time. I read about the professor that might just lead the way."

Jack, "Who is it?"

John, "Dr. Judith Curry. I read an article in the November, 2010 Scientific American, and I believe she has the guts that we need. Excuse the expression.

Jack do you know there are different political sides to the issues around climate change? Have you read about the Intergovernmental Panel on Climate Change?"

Jack, "No I thought science is science."

John, "You're just starting your homework. Read this article and then I have a list of periodicals and books for you to start on. You're going to have to learn how to tap dance."

Jack, "I just want to design and paint on ice."

John, "Don't worry, my brother Spike is the maintenance foreman at the university and he's won over a very prominent mathematics professor, whom I believe came from the Georgia Institute of Technology. Spike is a wiz at fixing and building computer capacity and he's solved a few problems for their department. Leave it with me.

In the meantime I think you should come and give a few guest lectures to my students and we can start to build your resume for the big day."

Jack, "I don't know how to teach."

John, "Young professors don't have a clue and they learn on the job. The only real teachers we have in my opinion are elementary teachers in the primary grades."

Jack made the T sign for time out and they both knew this was their new inside joke-John not getting tenure for having so much attitude and expressing it. Jack would be giving the T sign many more times.

"Jack, I have to run some numbers, but my buddy was called away by a death in the family. I won't be able to verify my calculations until we can sit down and go over all the figures. I think you're dead right about the loss of forests and their importance as a water sink. GVWS will be a new acronym."

"I think that GVWS could explain in detail a good part of the historic rise in the ocean levels. The forests also purify water. The groundwater consumed and stored in our trees eventually evaporates as pure clean moisture. If you calculate the surface area of the planet paved or cemented over, plus the footprint of all the buildings in the world and the loss of the forest mattress, you could conservatively calculate that the oceans rose about 6 to 8 inches. Also, don't forget to add the loss in the water table. I get upset thinking about it. We'll have to sit down and get a defensible number. I think another inch or two is conservative."

Jack, "Defensible number, what do you mean, defensible?"

John, "The scientific community has to know where you got your numbers, how they were calculated, and what assumptions you made. Then they will criticize every slip-up to death."

"Jack, the intellectual community is a very narrow community. You must follow the traditional way of thinking and doing things. The minute you venture off the path you're upsetting the status quo. All vested interests are protected at all costs and that's why the slightest slip-up will be blown out of proportion and a campaign to discredit you will ensue. Many people in this community do not have the conviction of their own intelligence. They need and seek the approval of their peers and only when they get the approval do they know that they are correct."

Jack, "Why do you want to be part of this group?"

John, "Believe me it's more rational than the other side. One more thing. My experience is just my experience. When you bombard people with too many ideas they close down and can't process the information. GVWS is a huge topic and it's worthy of its own time to percolate through the scientific and environmental community. If you also introduce the world wide depletion of ground water, it will overwhelm not only the scientific community but the general public as well. I think the best strategy is to say nothing about the depletion of the groundwater tables for a few months, maybe a year. There's no question when the records are dug out and examined the depletion will explain a major part of the rising sea levels.

I have a close friend that is a trapped civil servant. Veronica was a rising star and was moving up the ladder until one day she put in an improvement on how to make her operation more efficient. The idea was accepted and her division was able to save thirty percent of their budget. Five staff were not needed and transferred to another department. What Veronica didn't realize was that her direct boss would be scaled down on the salary grid as he now had fewer reporting staff. Veronica now spends her days keeping out of the way. She would love a new project. She's in the right department and has all the contacts to investigate wells, ground-water and what is happening and not happening across the country. If you are interested I'll set up a meeting and you can outline what research you need."

John, "Here's your chance to change that, for at least one person and maybe a few members of Veronica's department. You'll get a warm reception as they want something meaningful to do. Jack, the really important issue is you can't overwhelm people with so many concepts and ideas, they'll stick their heads in the sand."

Jack, "Can we make that, stick their heads down a well and watch the water level go down?"

John, "Look at the Oak Ridges Moraine project in your area. How long did it take that lobby group to pound information into politicians to stop the destruction of that huge water sink? The only thing politicians know is how to get elected. They will only commit to something when they can be assured there is a crowd supporting that decision. Get the herd of buffalo moving and they'll put up the road signs and take credit for their forward green thinking. Corporations are very much the same. The CEO is focused on the year-end bonus and the stock options waiting to be sold or cashed in. The environment and thinking green is just one of many tools sitting in the marketing package to be cashed in if it will increase profits. You can bet your bottom dollar that if it does not increase sales or bottom line profits you can whistle in the wind.

You've heard of the great marketing term called shareholder value? Sounds much better than, "we are running this organization to increase profits for the owners. Shareholder value gives the corporation licence to lay off people, close plants, screw thy neighbour. Corporations hire experts to spin not only the stories but the words."

Jack, "You're probably right. You know the Ontario government is still subsidizing farmers to drain wet areas on their farms. How counterproductive is that? In the city of Peterborough, they have had two years of really bad flooding. Guess what the city has done? They've filled in a large wetland south of the city and built a tourist information booth on part of the new found property. Get this-they sold some of this property for another coffee shop. Cups and lids will soon be filling up the drainage ditches.

Costco, a USA multinational, bought a piece of the land, or should I say wetland, to build their store and large parking lot. What are their corporate environmental standards? Filling in wetlands is part of Peterborough's history. George Street south of the Holiday Inn was all wetlands years ago. It doesn't look that way today. The old flood plain has many houses waiting for that once in a century disaster."

John noted:

WETLANDS FILLED DISPLACE WATER TO THE OCEAN

Jack, "Look at Toronto, the lakeshore was at Front Street. That's why it is called Front Street. They should rename it Back Street. How much fill has been dumped into Lake Ontario? How much water did the city displace into the oceans? This has happened around the world with the majority of coastal and inland cities. When property goes up in value it is then cheaper to fill and build into the wetlands or water. Good old Adam Smith's invisible hand at work again."

John, "Give me some good news."

Jack, "Get ready for this one. A group of guys noticed years ago that the number of ducks was declining. It is pretty hard to go hunting and shoot ducks if they are all gone. One and one make two and the boys soon figured out if the wetlands keep disappearing the ducks will have no home to raise their families. Ducks Unlimited was born and it has protected thousands of acres of wetland. This is the most successful environmental group that I know. The members of Ducks Unlimited, and yes there are plenty of women in the movement, fundraise and put their money where their mouths are. We owe a lot to Ducks Unlimited for their early intervention and protection of wetlands."

John noted:

Check out Ducks Unlimited, see Jean on staff - duck hunter

Jack, "John I am not a politician, this is out of my league."

John, "No, you're just in a different league. When your wife dresses up in her favourite dress and asks for your honest opinion, what do you say? That's being political.

Think about it for a moment. The industry has spent years drilling for water. Now you're saying they have to put water back into the water table. They'll come up with a thousand reasons and safeguards as to why you can't do it. You have to give them an economic reason to climb on board. Once you mention more holes have to be drilled, more studies have to be done, more research and special equipment is needed, you'll get their economic attention. Find out the employment and economic interest each group will exploit and once you've dragged them to the bank with their fists full of money, they'll support your views. Self interest always overrides community interests.

The fastest way to catch the attention of bureaucrats and bring them on board is to mention that they'll have to increase their departments for this important project. You see, as their department grows, they get bumped automatically into a higher pay bracket. Private business cuts costs while the goal of government employees is to increase their budgets and increase their staff."

Jack, "It is a good thing you are not opinionated."

They both laughed together.

Jack, "Holland built dykes and pumped out water to reclaim land. Holland Marsh in Ontario did the same thing on a much smaller scale. All that water went back to the oceans. How many other areas around the world have done the same? Let's talk about the state of Florida and how much water they've drained into the oceans. The old invisible hand of Adam Smith has cuffed us on the back of the head; almost like being back in school in the old days."

John noted:

Arizona, Florida how much water have they dumped back into the oceans?

Jack, "Parts of Florida are now two or three feet below sea level. They've pumped out so much ground water over the years that sink holes are developing and swallowing roads and buildings. There are insurance companies and construction companies that are focusing on this issue. Business is booming as Floridians are still pumping out water. Where do they think all that water is going? They've not asked the question, let alone not caring that they are part of the problem.

Rivers around the world are carrying sand and silt to the oceans, seas, and lakes and they're filling up one particle at a time. These natural deltas are lifting sea levels every day. We should be removing these deltas gradually and replacing the top soil back on to farm land.

One more thing. We used to build houses, barns and commercial buildings out of logs, wood siding and wood shingles. These wood products hold about 15 to 20 % moisture. Today subdivision after subdivision is built with plastic, vinyl, stone or brick siding with asphalt shingles. They hold no moisture.

Office towers and commercial buildings are glass, steel or concrete and they hold no moisture. We have made design decisions around insurance companies and short term economics." Look at the United Nations building that houses the group concerned about our loss of forests. They are in a glass and steel tower. They should at least have wood siding or porous siding to hold water. We have to rethink how we are building."

If buildings were made out of wood or at least the outside siding was made out of wood how much water would be held in buildings alone?"

John, "That calculation would make a great assignment topic for my new post graduate class. But John, wouldn't you be criticized if you housed yourself in a building from the forest you are protecting?"

Jack, Trees are a crop and they must be harvested wisely and used for our homes and offices. Trees are a replenishing resource. We have to monitor and manage the water capacity of our forests never taking out more water than is being grown each year by the forest.

Jack, "Air conditioners around the world remove the moisture from the air and ninety percent of the time the drain is running into the sewer system and off to the ocean. Homes, schools, commercial buildings, and office towers that have air conditioning all take moisture out of the air and cut the trippage short. "Dehumidifiers are connected to the sewer system and drain back to the ocean. Should we not collect that water and at least flush it down the toilet once?"

John: "Jack, I've got to run soon as one of the profs in the department has left on maternity leave. No one knew she was seeing anyone, let alone married and pregnant. I have a full time job for the rest of the year plus my two evening classes. Don't worry, we will talk, and Jack, just imagine having over three hundred and thirty undergrads and seventy-five post grads doing assignments on topics you want them to research."

With a wink, he said, "What topics will you be picking for your new free research team?"

Jack would soon learn how many professors got their free research done to survive the "publish or perish" rule of university culture. Jack would produce a research list that would choke a horse. Changing the structure of water, and creating different hardness levels at various temperatures now could be done in university laboratories.

Jack was always looking to the future.

Once Jack saw that newspapers and magazines were switching to vegetable dyes he knew the days of oil and acrylic based paints on ice were numbered. Jack was already using milk and vegetable based pigments but now there would be more in the market place and he would not have to make all of his own.

Jack: "John, look at this article from the Canadian Geographic, October 29, 2010 issue. I quote," John Pomeroy notes that thinning and partial clearing of trees can theoretically double available runoff, and these techniques are used in the Upper Colorado River Basin to increase river flow.

John, these are experts and they are reducing the water sink on purpose. Cutting down trees so snow can melt faster and run off? We need some big people, fast, to stop bad practices around the world. I can't believe the Canadian Geographic can be so irresponsible and so ill informed about our climate and vegetation as to publish an article like this."

John, "Jack, have you ever thought of doing a thesis for a doctorate degree?"

Jack answered honestly. "John, I'm a grade twelve graduate of Norwood District High School, I never finished grade thirteen.

Change can happen. You know all it would take is one municipality to close all drive-throughs and take-out services. The news media would pick up the story and run with it. There should be a deposit on all take-out containers and ninety-nine percent of the problem could be solved. John, they are not take-out windows, they are throw-out windows. When my wife and I go to the cottage we take a walk every morning after breakfast. We take a bag and a stick with a sharp nail on the end and we fill it with garbage. Every morning during the week we find fast food litter in the same location. On the weekends there's a different pattern. Why can't a company set an industry example and put a deposit on their products? Eventually they'll be dragged by another government rule, kicking and screaming into environmental responsibility. They'll then put a positive corporate spin on it. "Look-we are champions of the environment and look how good we are." I would solve the problem. A new cup with a plastic handcuff attached to it and to your wrist. The only way to get it off is to put your hand over a recycle container."

Jack, "I am not sure who came up with Timbleweed but it stuck and the ditches are full of it and plastic bottles because there is no deposit on them. Other provinces have a deposit and they do not have this problem."

John, "Just to review, so I have not missed it, ocean levels are overflowing due to:

- Loss of trees as they are the GVWS of the world.

- Loss of wet lands.

- Drainage of the underground watertables.

- Cities, towns, and villages around the world filling in their waterfronts.

- Reclaiming land like Holland and on a much smaller scale, Ontario's Holland Marsh

- Farmers, many times with government grants, putting in drainage systems to drain their fields.

- Concrete and asphalt being the two biggest culprits as they dry out the natural sponge material of the planet.

- Gravel holding water. We turn it into concrete and put it down for a road or up for a building. We've lost the sponge.

The world makes five billion cubic yards of concrete every year and it is increasing. It's the largest man-made substance in the world. We continue depleting the sponge every year as we continue to build.

The only natural way oceans are overflowing is the silt being washed into the bays, deltas, and river mouths filling up the reservoirs."

John continued, "Jack, have you done any calculations on the amount of water each has dumped into the oceans or a rough percentage of the problem?"

Jack: "I am neither a mathematician nor a scientist but I've worked out some rough numbers. There is a formula for calculating the surface of a ball and our planet is just about a perfect ball and close enough for me. You read many sup- posedly well researched articles stating that if the polar ice caps melt, the oceans will rise five, ten-and one article said over one hundred-feet. Impossible. Sadly they're not called on it. I may be repeating myself but if you think of ice cubes in a drink. The ice melts and the water does not flow over the rim of the glass. Only melting ice on land will increase the water level.

Every country in the world that had trees has cut them down. There are only a few pockets in the world that still have not made their first cut. I have calculated that if we put back the world's trees in every spot we can lower the oceans about three or four inches easily.

Loss of wet lands around the world accounts for a rise of three to four inches, or about eight centimetres. If it was not for Ducks Unlimited, we would be in bigger trouble. They put their money where their mouth is. Ducks Unlimited is the only privately funded organization I know of that is protecting wetlands. We need to support this type of initiative.

Oh, checking my notes, I just about forgot to mention the peat and peat moss harvest around the world. Nature's water sponges are drained, dug up and peat is burnt for fuel. Peat moss is spread elsewhere and eventually dries and is blown away. We are talking billions of tons per year. How much water is dumped into the oceans by draining and burning these huge natural water sponges not to mention the loss of evaporation surface to allow our planet to keep cool?

Loss of underground water is the biggest, the easiest and fastest to fix. I estimate that if every country started to put back their ground water we would lower the oceans three to four inches, say nine centimetres.

New concrete must not be made from gravel. It should only be made from reclaimed concrete or crushed rock. The water sponge effect would be minimized. Gravel and sand should not be used. It is a precious sponge and should be preserved.

Silt being carried into the reservoirs of the world should be removed on a systematic basis and put back on the land. The world needs this silt to grow crops."

John noted:

Hydration check it out.

The living sponge: Surface sponge material (SSM)

Jack continued, "John, if you look at the part of earth that is above water, it's roughly close to twenty-nine percent of the surface. We have, as intelligent individuals, a choice on how to use that twenty-nine percent as most of us will drown in the other seventy-one percent as we do not have movie star gills. If we continue what we are doing, we will all die together and that is a good thing, because we all have to make big decisions. No one is off the boat. I call the boat seven billion rats floating on a large piece of cheese and they are eating it. We all go together rich and poor, informed and ignorant. Ignorant is not a sarcastic word. It just means not informed."

John is keen to establish priorities with Jack.

John, "We have to focus on one thing at a time. You'll overwhelm people and they'll crawl back into their shell and do nothing. You eat an elephant one fork at a time and you introduce change one step at a time. Expect to go two steps forward and then one back. Two steps forward and one back.

Jack let's look at concrete:

Gravel beds, eskers, drumlins, and moraines are full of gravel and covered with a layer of top soil and growing trees. Areas that have had the trees cut down and are growing crops are still acting like Earth's sponge. Huge volumes of water are stored in these land forms. Not only is the natural rainfall and snowmelt filtered and purified by the gravel, it is stored to support the vegetation on the surface.

How much of the Earth's sponge has been turned into dry concrete around the world?

When we take gravel, the earths' natural water sponge and turn it into concrete and construct roads and buildings, you're ringing out the water and sending it back to the oceans. Crushing stone and then making concrete has less impact on the earth's water sponge.

Jack, once the information is out; the corporations that own gravel pits will buy consultants and university experts to give independent evaluations. Get prepared for a storm of ridicule and slander. Do you have a tough skin to withstand the storm?"

Jack, if you were to pick one country, one province or one area in the world what would be the most environmentally unfriendly spot in the world?

Jack, "Most people would think of China and the developing nations. I would pick the Cayman Islands as the per capita biggest polluters on the planet."

John sat with a blank, stunned look on his face and his body communicated disbelief in the answer, Cayman Islands? Slowly out of his mouth the words dropped out. "Cayman Islands?"

Jack, "Here is a sand bar in the Caribbean that has one of the highest per capita incomes of the area. They use the cheapest diesel fuel they can buy to run their generation plant to produce electricity. The sun shines almost every day of the year go figure that one out.

There are no emission tests on any vehicle on the island. The right hand drive system is convenient. They import all the old vehicles from Japan that have gone by the date standards or have not passed the Japanese emission standards.

They recycle absolutely nothing. Not one beer bottle, newspaper or plastic bag or plastic bottle. On a small island they land fill everything. They tried a recycle program a few years ago and they could not keep up with the amount of material people wanted to recycle so they cancelled the program.

All the fresh water is produced from the ocean through a desalination plant. No rain water or recycle water program is in place. Black water is piped out into the ocean through a large long pipe. This is one of the most pristine areas in the world for scuba diving.

Farming is minimal. Ship loads from Florida supply the population and the tourists. No composting is done of any vegetation.

John, I can go on but you get the picture. Small tropical island, tax free haven and the winds blow everything out to sea.

Oh yeah, they charge cruise ships to anchor in their harbour and fine the ships heavily if they discharge their black water tanks in the harbour.

John, your comments about dollar votes. If tourists knew this, would they support this destruction of our planet?"

John, "Jack, let me think about a strategy and at our next meeting I'll run a couple of ideas by you."

John's biggest challenge was to coach his new friend, Jack Wilson on how to introduce change to a very sceptical population.

John thought to himself. How can I take this man's passion and infuse it into some of my students?

CHAPTER TWO

Jack Goes To University

J ack looked down and saw the small microphone under his chin. Without thinking he touched it again to hear if it was working. The chalk box was full, the boards were clean. His hands were sweating. One part of his brain was saying, What have I got myself into? The other side of his brain could hardly wait for the students to arrive. He looked over at John sitting at the side of the lecture hall, head bent over, busy marking papers.

The first group of students started to pour into the auditorium.

John Henry had asked Jack if he would consider being a guest speaker for his first year general science course. John had a few reasons for asking. The first was that the young students would benefit from hearing a person like Jack expound on topics and give theories that they were not going to find in any textbook or journal. This was raw, unedited knowledge ready to be vetted. The second reason was it would be good for Jack to have a critical audience and to get their reaction to his thoughts. John had the feeling that Jack's future would be on stage in many cities, slowly changing the direction of the environmental community. Jack would also have an opportunity to meet the students that would be researching the topics Jack had chosen.

Jack, "You know John, I talk with my hands. At our monthly meetings we have a wall of blackboards that I salvaged when they were tearing down the old Newmarket High School. Real black slate. The school board officials considered slate to be unsafe for schools as slate is a cut and polished stone, as you likely know. One slate somewhere on the planet was struck and broke. The stone fell down and nearly hit a student. I think it was the green board suppliers that threw the sledge hammer. I'm a doodler and I talk and draw at the same time. There is nothing like black slate with chalk to write and draw on. The Italian and French students won't need an interpreter for my hand language.

Do you have a room with real blackboards at the front?"

John, "My lab room has those green painted metal boards on three sides and windows on the remaining wall. It is not large enough. I'll book the lecture theatre in the next building. This Saturday morning it's probably available as it's in use every weekday. Most evenings are booked for night classes. Let's go and have a look at the hall and we can see if it has the boards you need."

The old lecture theatre was sloped towards the front with a good view from every seat. Sure enough, across the complete front was a row of real blackboards from one side wall to the other, and the boards were very tall.

Jack, "This should be just fine. We can move the pulpit out of the way."

John, "That is a lectern."

Jack, "Looks like a Baptist church pulpit to me. Isn't this room a bit big for your class?"

John, "No, they load up the numbers in the first year so we can have smaller class sizes in years two and three. Graduate school classes have even smaller numbers. My department head and dean were ticked off with me when I told them they were wrong. You need smaller classes the first year so you can reduce the high failure rate, then each year after, you can increase the number of students as they have adjusted and become more independent. That fell on deaf ears, as guess who teaches the smaller graduate classes?"

John, "First year students are cannon fodder. They pay good money, subsidize the senior students, and are blown away. Universities that catch on to this game grow."

Jack, "I can see why you are a popular guy on campus. John, I'll bring my chalk from home and practice Friday night if we can get the room after the last evening class for an hour."

John: "There are no classes Friday evening. Pubs replace lecture halls. Many times the discussions are better than the lectures here. Professors have learned to lobby not to have a Friday evening class, as attendance would be minimal."

The lecture hall was available and Jack came back to Ottawa Friday evening and was left to practice. John was preparing lectures, then he had marking. They agreed to go out for wings and beer after ten. John called home to see if his wife, Helen, could get a sitter and join them. It was Friday night and both needed a break. Jack was very much at home with all that blackboard space.

John brought his Sunlight soap and bucket, and when he was finished he washed the slate boards.

By nine-thirty Saturday morning the custodian had finished cleaning the lecture theatre. John and Jack walked in and turned up the front lights and moved the lectern.

Jack was very nervous as he was used to talking to only a few employees or arena staff that would number at the most a half dozen. He decided that the lecture theatre was just a bigger Zamboni room. He convinced himself, "I'm here to sketch. I have drawn on ice many times before. This will be much warmer and with chalk.

The first year students started to wander in and most were present by ten. John counted the audience and thought one hundred and thirty-three out of one hundred and sixty-eight was a very good turnout for a Saturday morning. Friday night is one of the two big party nights and the lack of sleep and dead brain cells had more than a few students on auto pilot.

Jack sat quietly at one side of the lecture theatre checking his notes, then he picked up the smell of university students. Elementary, secondary schools, and universities had their own particular odour. This theatre had a fresh smell with hints of rain soaked outerwear, and without the popcorn.

John introduced Jack Wilson.

John: "Today you have an opportunity to meet someone special. You've heard the expression about thinking outside the box. Jack Wilson not only thinks out of the box, he is expanding the ball of elastic bands academics keep secure in their cupboard drawers. My friend Jack needs help in researching different topics and you will all be helping. Your essay topics for this course have been selected by Jack and myself. Your research and knowledge will be valued. I will grade the essays as required by the university, but Jack will also be poring over every word you write. Jack Wilson is an ice man. He owns and operates a company that paints on ice. Painted Ice Inc. is a company with its headquarters in Richmond Hill, Ontario. The staff paints arenas, curling rinks, and other large ice surfaces all over North America and parts of Europe. You probably all have seen some of Jack's work before today. I believe you'll find his insights and observations refreshing. Please welcome Jack."

A courtesy clap of hands woke up a few students in the back row, and a nod of thanks was sent out to the crowd from Professor John Henry.

Jack had on his pouch and it was full of coloured chalk of various sizes and one clean brush. He walked briskly to centre stage and turned to the audience. Touching his headset he asked the students in the back row if the volume was high enough. Heads nodded and Jack nodded back.

Jack immediately caught the attention of his audience. His speed of entry on stage, his crystal clear voice, and his body language commanded respect from the start. Jack was very agile for a man over two hundred and forty pounds and six foot five. His hands were very large and ten fingers looked almost swollen. The cold ice over the years had grown his paws big. Because Jack's body build was in proportion and there was no lectern to give size reference, students would not realize how large he was until they walked up to meet him after his first class. Most of the students had at least one professor who was at first hard to understand. They had to cut through the accent and adjust their listening to know what was going on. A few students at the back wanted to sleep off the good time they had the night before as they were trying to function on a couple hours of sleep. Quiet is what the brain wanted.

THE STAGE WAS SET

Jack spoke softly into the microphone. "Good morning. I prefer to be called Jack; my dad is Mr. Wilson. If you have a question please give me your name first. Feel comfortable to interrupt.

I need you to work hard and accurately on your assignments. I need your research as I don't have the time to do all that has to be done."

The board was jet black and waiting. In the top left corner Jack wrote, "1493 To Today."

Jack: "School is out." A few students smiled.

Jack started at the left side of the room and drew a white chalk line completely across the front of the class to the other side. It was a raised arch and he said into his headpiece microphone, "This is the curvature of the earth from Victoria to St John's. It is slightly less than scale." Then Jack quickly drew another line almost on top of the first, and he said, "That is our topsoil in Canada. Our good stuff got pushed south in the ice ages. Our farmers work with six to eight inches."

Jack drew in the west coast valleys and mountain ranges, the low valleys of the prairies and he worked his way across Canada to the mountain ridges in Quebec, then on out to the Atlantic provinces.

Jack was gaining confidence. He was pretending to be talking to the rink rats in the, oh-so-many, arenas he had lived in all his life. The students sat transfixed. It was as if they had slipped into a studio or were standing behind a street artist and watching him sketch. Jack was a master and he worked quickly. Years of working on a cold ice surface had taught Jack to keep moving, keep warm, and get the large job done.

The young students were seeing their country from a new perspective. Students that had not travelled across Canada were seeing it for the first time. It was like Canada was sliced open and on the operating table. Jack was working fast, as he had a large country to cover. Students drifting in late sat down and became glued to the board, they wondered what they had missed and were upset with themselves that they had not come on time. It was a bit like getting into an action movie late. The first few scenes set up the suspense. They saw the date in the top left corner: "1493 To Today."

Trees, forests, prairie grass, lakes, rivers, swamps, and mountain ranges; they were drawn showing how the earth's crust lifted up here, folded over there. Valleys scraped out by the glaciers and granite bedrock grooved in the direction of the last glacier were now understandable and ruggedly beautiful. The Niagara escarpment with Niagara Falls was captured in one corner. Students could see how the ground fell and the water flowed over and down into the valley floor.

Once the cross-section of the country was in place and the large city names were identified, no one questioned that Norwood was between Toronto and Ottawa. John Henry smiled and the students wondered if they had missed that large city Norwood.

Jack took a short break from his sketching and said, "Now we can add the weather. If we don't, Canadians would have nothing to talk about."

Students were warming up to Jack.

"Picture the earth as a round ball rotating east to west and the atmosphere above not able to keep up. This is an abbreviated version of the Coriolis effect. I will put that name up here so the non geography students can read about it. The atmosphere is trying to move with the earth but it does not have the same mass or density to hang in there. Think of our planet Earth as a ball bearing covered with a lubricant called atmosphere. This lubricant allows Earth to spin friction free on its axis. At the equator the earth's spin is close to one thousand miles per hour, while this ball bearing is pulled by gravity around the sun every 365¼ days. If our planet loses this lubricant, friction will very slowly stop the spin. Earth will then be extremely hot on one side and very cold on the other. We are on the surface of the earth sliding underneath this lubricant."

A couple of geography majors looked at each other and the knowing look brought a smile. This lubricant concept was definitely not in the books they were reading. This guy, Jack Wilson, did not use the terminology of a professor but they understood everything he was saying.

Jack, "The westerly winds suck up moisture as they move across the ocean and drop it on Vancouver Island and on the west side of the Rockies."

Clouds were drawn in and rain drops were shown falling onto the huge west coast fir trees. Water soaked into the ground and extra surface water trickled into the streams into the rivers, back into the ocean. Jack was so absorbed in his board work that he didn't hear or see the students from the back moving up to the vacant front seats.

A large west coast Douglas fir was sketched into the upper left corner of the board. Jack did not have to say a word. The students watched as Jack drew rain dropping from needle to needle, while some splashes made their way to the scaly bark on the trunk of the tree. Gravity was pulling the blue droplets all the way down to ground level and saturating the soil and rotten logs.

Jack, "Do you know that in those old forests it had to drizzle almost a half hour before a drop would get to the forest floor? The steady rain soaks and it took a long drizzle or many quick rains for water to get down to the forest floor. Because of the shade and the fact that there is practically no breeze at ground level, the organic layer stayed wet for a long time. Rain had time to soak in.

How much water did our original forests hold? This would be an interesting essay topic for a student that has a way with numbers and is good at estimating and projecting. Excess rain and melting snow re-fills the underground water table and extra surface water runs into the creeks, streams, and rivers.

Two words you will hear me say." And Jack wrote on the top of the centre board, "Trippage Slippage".

"How many times does that water cycle turn over before the west wind pushes the moisture above the Rockies towards the east?

Trippage, I call it trippage. How many times does the same water molecule drop on Canada, soak in, gets evaporated, frozen, sublimated and falls again on Canadian soil? From side to side, bottom to top, it is not only the amount of precipitation that counts, it is the trippage that is most important.

If the trippage is only one turnover we are in trouble. Cut down our forests and you cut down the trippage. Increasing the trippage not only is good for plants, but it cools the planet.

How many water turns-what is the trippage-from our west coast to the east coast? What is our trippage from the North Pole to the forty-ninth parallel?

What is our trippage when a gulf storm roars up through the United States bringing all that water moisture, and I might add pollution, from the industrial heartland of the USA? On our east coast when a storm blows in and drenches or buries the folks down there, bys the way, they call it a Nor'easter."

The same two geography students looked at each other and they both wrote down trippage and added a big question mark.

Jack adjusted his head set and continued: "The forests determine the number of turns and the turns dictate our rainfall. If you think of all the people up river flushing their toilets so the people downstream can have drinking water, you'll get the idea of water cycle turns or trippage through the system. Think of the Great Lakes and the basin it drains. How many times has the water been flushed down a toilet before it gets to Quebec City? It's no wonder our French neighbours downstream prefer wine over water."

The audience broke out into laughter.

John Henry was starting to relax. It's always a risk when you invite someone to speak to your class. You're never sure if it is going to be a washout or a success.

Jack continued speaking into his microphone and walking and looking into the eyes of as many students as he could. "They don't walk the talk either. The cities along the St. Lawrence are still running untreated sewage into the Atlantic Ocean. You wonder how the fish out there can survive and by the way, we're still eating that fresh wild seafood."

The audience was warming to Jack's sarcastic sense of humour.

"Have we factored into our climate predicting models the number of water turns, or saying it a different way, does the computer model build in a trippage number? When the hay, corn, and soybean harvest are on, how many million acres of crops are giving up their water? When the leaves and the two year old needles drop in the fall, do we factor in the volume of moisture released into the atmosphere. No wonder we have fall rains.

Slippage is when we don't get to use the moisture. In the winter when it snows the snow that falls in the forest areas is protected, unlike the open field snow. In the open fields the sun sublimates much of the snow and the moisture goes down wind to fall as snow, ice, or rain. This process can continue until the moisture eventually gets dumped back into the ocean. Snow that falls in a forest is a different thing. There is far less snow being sublimated because the trees block the sun. If we get a lot of snow the frost in the ground comes out due to the insulation of the thick white blanket and much of the moisture goes into the ground and the water table. Folks, there is no table; it is just a system of veins and pockets all connected together. The excess water that does not get soaked up by the vegetation, ground cover, and tree roots, runs off into streams, rivers, and lakes. The moisture is stored temporarily."

The diagrams are drawn as fast as Jack is talking. Sometimes one or two lines are all that is needed to project the image that Jack is talking about.

Jack draws a large lake and shows how water is evaporating and forming clouds.

Jack continues as he draws. "Along the rivers and at the end of lakes we have dams to hold the water back. Holding and ponding the water allows for more evaporation and also the generation of electricity. The other important feature is the dams and canal systems, like the great lakes, allow for shipping. Evaporation is the primary way our earth cools itself. When we cut our forests we reduce not only the evaporation through the leaves but we also lower the trippage number. Our individual actions are heating up the planet. Reduce the opportunity for evaporation and we prevent the planet from reducing the heat of the sun."

Jack turns to the students with chalk dust on his hands and clothes and says. "You know about global warming or climate change? I think if you put the trees back and stopped burning all the fossil fuels and stopped burning all the electricity the meters would show that the world has cooled down. In the spring alone, the snow and ice in the forest hangs around a good two weeks longer than the open farm fields. Farmers can be working their crop fields and over the fence in the woods they'll see snow nestled in around the tree trunks. Two weeks out of fifty-two is four percent of the year that we have snow and ice longer on the ground in the woods. Cutting and clearing our trees is preventing the planet from staying cool naturally. Humans are stopping the world from keeping cool."

The two geography majors looked at each other and one whispered. "We're looking for ways the world is heating up. Let's add slippage to the list. I think we are stopping the planet from cooling down."

Jack then draws the American West and sketches in California and Arizona.

"You can see what is happening here. Millions of water users are living in this area and there is no annual refilling of the water supply. Russia and parts of Europe have the same problem. The ground level in Mexico City has dropped over twenty-five feet in places. Sink holes are common not only in Florida but in many spots all over the world. Suck out the water, don't put it back and you will create cavities. Gravity will create the holes."

Jack draws a side view of an ice road on a typical lake in Northern Ontario during the winter. Three feet of ice on top of the water and a tractor trailer loaded with supplies is heading north to stock up the warehouses for the summer. The forty ton load moves across the surface and continues on. Jack then takes his brush and carefully removes two feet of water below the ice surface. The next truck doesn't make it.

Jack does not have to explain to the audience. Sink holes in Florida and everywhere around the world are caused by removing the water, plain and simple.

A voice comes from the audience. "Friday Hope is the name, Professor, you can call that sippage. They are only going to get it once and it's gone."

Without batting an eye, just giving a wink, Jack handed the chalk to Friday and motioned for him to write it on the board.

SHADE, EVAPORATION, ALTITUDE (SEA)

On the board trippage, slippage and sippage stood out. Jack pointed to the three words and he said, "Add evaporation to that and you have started on the path to global warming or global cooling. The planet only has three ways to cool itself: shade, evaporation and altitude (SEA). That is it folks, there is no other way to cool our planet. We are systematically preventing our planet from cooling itself. We also have found ways to keep on burning and adding to the heating cycle. We, seven billion humans, are stopping the planet from cooling. By the way, the A stands for altitude. When water moisture rises and forms clouds the higher altitude cools the molecules; then the precipitation that falls not only cleans the air but cools the surface of the planet."

The sketch clearly showed the cooling water falling on the earth.

"Trippage determines the amount of evaporation and we're reducing it. By cutting down our forests we're also reducing the shade. The earth's rotation causes the night shade and the tilt with the rotation around the sun causes the four seasons."

Jack paused for a moment and looked at the audience. After a moment of complete quiet as the information was sinking into the young minds, Jack broke the silence. "Ladies and gentlemen, this is not rocket science. This is a fact, and we have to reverse the damage that we've done to our planet or we're going to pay the price over and over again."

The two geography majors no longer had smirks on their faces. They noted SEA, shade, evaporation and altitude. This simple theory on cooling was not in their curriculum. In fact they had never heard of a cooling theory. Bernie whispered to his friend, "This guy not only thinks out of the box he's looking the other way."

Jack, "When two carpenter ants chew into a pine log they do not realize that breeding, eating, creating tunnels, building homes, and doing what is in their self interest will eventually result in thousands and thousands of ants which will consume their log. They consume their home, their protection. Lucky for ants they can move onto another log or tree. We have only one Earth and we are not much smarter than those ants."

The permanent glaciers were sketched on the board and Jack printed June at the top. Snow was falling. Jack put a dot on the glacier and said, "That is where I fell, coming down the Saddle at Whistler. They cut me off our Olympic team last year, and my friends said I should take up face planting."

The audience that had been to the west skiing, laughed. The rest joined in.

Jack coughed a dry cough and tried to continue talking. He stopped briefly and looked around. A student at the front of the class handed Jack a fresh bottle of water. It was Friday Hope. This gesture registered with all the students and showed that Jack was just a bit inexperienced in public speaking.

Jack was drawing a dry breeze, compared to the west of the Rockies. It was sliding down the east side of the Rockies out onto the prairies. The moisture content was much lower than on the west side but there was enough water vapour to have a little rain. A cool air mass was flowing down from the north and the line of turbulence was bringing rain to the grain basket of Canada. The water droplets were being pushed back up by a colder heavier air mass pushing from behind.

Jack: "The water droplets are growing in size and freezing. It will be hail today for a few farmers below. Look now the grain is beaten flat onto the soil and no crop will be harvested this year.

This is the part of Canada where we grow the wheat for Italian pasta and barley for Scotland to produce their famous single malt scotch. Oh don't forget you need a slice of bread for breakfast toast and there is nothing that compares to a Montreal bagel. Grown in the west should be stamped on a lot of products not only in Canada but around the world."

Jack worked his way across the country drawing in storms, air pressure masses, and on the East coast he had a hurricane heading north just about to strike Halifax.

Will the westerlies push this hurricane east away from the land mass? Today I will brush it past and just dump rain and lots of it."

Two hours slid by very quickly and soon it was lunch time. John was even caught off guard as he hadn't looked at his watch and was unaware of the time.

Jack, "I have to take a break for lunch and I'll be back to finish this off at one o'clock."

John Henry took over the class and announced, "I know this is unscheduled but Jack will be completing this lecture at one o'clock if you have the time to come back. If not, have someone take notes for you. I will have handouts next class on the work you miss."

John, "Let's go over to the cafeteria and grab lunch."

Jack: "If you don't mind, I always pack my lunch as I never know where I will be when I get hungry and I want to change some of the board work before one."

Around twelve thirty, students started to drift back into the lecture hall. They sat quietly watching Jack rub out and make changes to the landscape. Oil wells, oil sands, factories, houses, highways, parking lots, airports, cut away pictures of storm sewers, mines, pipe lines, and underground services were boxed in across the top of the board in close proximity, as you would find them today.

At one o'clock Jack looked up. The lecture hall was full and students were sitting in the aisles. Jack looked around and said, "Friday, I hope you're here." The audience laughed. A large student in front spoke up, "Friday is the name and this is the second time I have ever wanted to sit up front."

Jack, winked. "Thanks for the water," he said.

Jack tapped the microphone below his chin and looked towards the back of the auditorium and got a wave of hands. He was ready to go.

"Time has passed and we're in the late sixties. It's July first."

Jack pulled out more chalk and went to work.

Jack started to talk into his microphone. "Let's work our way from west to east. We're going to have to erase most of the huge two and three hundred foot trees in the west and replace them with these fifty-footers. We'll keep this large one on the board and revisit it in a few minutes. Our trees today are like shrubs in comparison to our forests before the first cut.

In the valleys we'll wipe off all the trees and draw in twelve inch plants. Let's remove a few mountains for all the coal that has been extracted and shipped to overseas markets, especially China. You know they're not Canadian mountains. We sold them long ago. We needed foreign investment-that's for your political science and economics' professors to explain.

These west coast rivers were pure and clean but today we have to brown them down and remove most of the fish. Let's see, a dam here and a dam there. That will plug things up a bit. Moving to the prairies we'll remove those herds of bison as they were eating up all the saplings and rubbing the bark off the bigger trees. Let's just remove all the Indian villages and put in their place cattle ranches and large fields of grain.

This is where we seriously start punching holes in the earth's crust and extracting crude oil and natural gas. Not enough holes so we will dig some big pits for the tar sands projects. The open pits are only small test pits to become monsters once the price of crude increases. Please note I did not say the cost of extruding would force the price up. That is a question for your economics class. Not to worry, they are not Canadian. The holes and crude oil are owned by foreigners. Do you think the Middle Eastern countries are selfish for owning their own crude and raw resources? I guess they have not heard of the need for foreign investment."

The audience laughed.

"Now that we've sold off most of our raw resources, and cut down just about all of our forests, we will continue to over-farm the thin topsoil of the prairies. Roads and airports connect the cities together and let's not forget the railroads that were sold to the public to opening up the west. Maybe the favourable Crow's rate system should have been for passengers instead of trainloads of grain headed east for the export market.

Farmers in June are busy haying. They have to get their first crop dried and baled. They don't realize it but when they cut thousands of acres or thousands of hectares of hay, they're creating the rain. Over eighty percent of the hay that is cut is water. It wicks up into the air and forms clouds to rain on the farms downwind. If farmers worked together they could crop downwind first then move up wind on a timed cycle and most crops could be cut, dried, and baled without the inconvenience of rain. Rain would come after the baling to give the second crop a jump. Farmers are too busy looking after themselves to see the big picture. They're just one small invisible hand again."

A large laugh exploded from the audience.

Jack turned to the audience and said, "You know, most people are buried six feet under when they die but not farmers. Do you know why they are buried only 16 inches under the sod? So they can keep their hand out." The farm students had a quiet chuckle while many did not get it. Ralph was from a large dairy farm and his family had received frequent provincial grants. He remembered one large grant they received a few years ago to drain a twenty acre field next to the boundary road and five years later they got another government grant to return it to wetlands. His dad called it the government field. It didn't hurt that his dad was the warden and his dad's best friend was the government manager that allocated the government funds in their area of the province. His dad often said that the best thing he did was marry Ralph's mother. She was an English major at Guelph University where they met. She helped dad with his essays and dad made sure she had free farm food for the three girls that lived in the rented flat.

Few farmers had the time, knowledge, or ability to complete all the paperwork demanded by the government. That was just the start of the paperwork. Follow up paperwork, filling out forms, and surveys was just as important to the government handout or they would take the grant back. Our farm accountant said that Ralph's mom's time was worth more than $500.00 an hour.

∾

Ralph turned to Bernie, "I think this guy has been around."

∾

Jack carried on, with chalk in hand. "Let's look at Ontario. There we go, Sudbury and the cloud of acid rain blowing east and south. You wonder why the Muskoka lakes are so clean looking? Just a bit south of Ontario there is the Detroit River that had so many chemicals dumped into that waterway, it caught fire every once in a while. Fire fighters are standing by. The big auto companies are punching out more automobiles and industry is humming, producing many goods that the rest of the country needs.

We have to add smog clouds over Toronto. Gushing up north is the polluted industrial air from the Ohio Valley.

We'll get even with our neighbours down there. We'll send our geese south. Those geese are like my in-laws. Uninvited, they come and eat, don't stop talking, and dump all over the place." An outline of a goose was quickly drawn heading south.

Jack, "Our loons are smarter. They change colour and fly south in late fall and keep their mouths shut until they can return." More laughter.

Jack continued sketching and drawing but mainly rubbing out forests and vegetation as he worked his way across the country. On the east coast he wiped out the fisheries and the forests. Clear cutting was no sweat when you had a chalk brush in your hand.

Jack, "You know when I was a kid in grade eight we went on a class trip and took the Thousand Island cruise on the St. Lawrence River. We stopped at Bolt Castle. On board I had to go to the washroom. I noticed when I flushed, it went directly into the river. I mentioned this to my teacher, Mr. Fisher. He said, "Remember, don't flush when we are at the dock. Wait until the boat is moving." Mr. Fisher also explained that all the cities along the great lakes and rivers running into the ocean bury a pipe out as far as possible and run their sewers into the waterways. Most of our cities around the world, including Canada, are still running sewage straight into the oceans.

Over two-thirds of the world's population live along the coastline or on waterways. Things haven't changed much. If the so-called wealthy countries are not treating their blackwater you can imagine what the other countries in the world are flushing into the rivers, oceans, and seas.

Ocean ships and cities have dumped their garbage for years directly into the ocean. The barges that were pulled out from New York City were so large that bulldozers rode on top and were used to push the high piles overboard. We wonder why there is a collar of plastic floating in the ocean. Ocean cruise liners, commercial ships, and military ships all dump their sewage holding tanks at sea. You used to see signs on trains, "Don't Flush In The Station." Same at sea, don't flush the holding tank in the harbour. Cruise liners are just big floating toilets. When they board, passengers should have to wear buttons that read I don't give a shit. It's not the poor people that are taking vacations on those floating toilets. Sure, whales and all the live species in the ocean defecate but these cruise ships have thousands of passengers and thousands of crew members flushing every day. We wonder why we have a plastic bubble thousands of miles long and wide floating in the sea. Men and women flush a bunch of stuff down those toilets. Think of all the medication the drug industry pumps out for profit. It is urinated and defecated into our drinking water and flushed into our lakes, rivers, and oceans. No wonder some of the whales and porpoises beach themselves. Why do we flush all this crap into the oceans of the world then turn around and eat the catch of the day?

Jack paused for a moment caught his breath and continued: "I am off topic. Just one more point. If the coastal cities and harbours around the world are dumping and flushing directly into the oceans they sure do not have the facilities to unload and treat the blackwater from the ships docking in their harbours.

Let's go back to the west and look at the large Douglas fir tree."

Jack moved to the centre board, wiped it clean, and drew a huge tree.

Jack sketches as he talks, "It rains and the drops collect on the needles and bark. After thirty minutes the moisture finally hits the ground. The ground becomes saturated and the water continues down into the underground water-table."

The tree is being radiated with sunlight and photosynthesis is taking place. Notice the water in the root system is crawling back up the tree and the branches are getting longer and more needles are growing. The stem of the tree is getting larger and you can see stretch marks on the bark as the outer layer is stretching outward to allow for growth. If you see stretch marks on the stem of a tree you know it is growing. Most of you are from Ontario and you are all going to this university in Ontario. The next time you take a walk in the park look at the oak and maple tree trunks at eye level. The width of the stretch marks on the trees is a good indicator that they're healthy and growing. Oak stretch marks are light orange in colour, but remember my wife thinks I am colour-blind. Birch trees can't hide their growth as the bark has to peel off and fall so the tree can continue to grow through that water-proof girdle. Birch bark is the only bark that I know that is water-proof. Next time you're in Peterborough, Ontario, visit the canoe museum. It's world class and worth the visit just to see all the various birch bark canoes."

Jack drew stretch marks and peeled birch trees as he was talking.

"Here is what most people don't see." Jack paused and took out his large blue chalk.

"This Douglas fir is two hundred and fifty feet tall and it probably has three thousand board-feet of lumber in it. The branches would probably make three cords of wood. The needles alone would probably weigh close to a thousand pounds." As Jack was talking he was colouring in half of the stem blue, eighty percent of the branches blue, and the leaves ninety percent blue.

Jack turned to the students. He paused and said, "Trees are full of trapped water. They are the largest reserve of fresh water in the world. This one tree probably holds, well, you can figure that out. It probably holds close to three thousand pounds of water in the needles, branches, stem, and roots."

Ralph turned to Bernie, "Did I hear that correctly?"

"We Canadians and the rest of the world have cut down our trees not realizing they are holding our fresh water supply. Many people today say that our lakes and rivers are the largest fresh water reserves of the world. Well, it used to be our trees. Over seventy-five percent of the earth's land mass was treed and we've cut down over ninety percent of our trees. Trees are responsible for the water turns. Remember the term trippage. Trees are also busy cleaning the ground water sucked up by their root system. The leaves and needles allow this clean water to evaporate into the air for downwind users. How many times does the moisture, picked up from the Pacific Ocean, get consumed as it crosses our country? If it landed only once we would be in trouble. Can you calculate the number of water turns? The turns determine the amount of rain we get each season, each year. Our forests, our trees determine the number of turns. Sorry for repeating myself but it is so important. Water turns, called trippage, is our key."

The class was dead silent. Jack waited.

Jack, in a slow but very clear voice, "We have cut down our forest and destroyed our green vegetation water sink. Do you think our historic rise in ocean levels will match our destruction of our forests around the world?

We cut down our forests and we drown our coastal friends around the world."

Bernie turned to his buddy, Ralph, "I can't believe my ears. How could we have overlooked the water capacity of our forests?"

A young female student in the centre of the lecture hall raised her hand and Jack acknowledged her. "My name is Sonya Doherty. "Jack, can we reverse the mess we are in?"

Jack: "A very good question Sonya. The answer is a resounding yes! We will look at some solutions next week. Remember we got into this mess one bit at a time and we'll have to break some bad habits to get out of it. Here is one last example for today. The next time you purchase a bottle of wine, a case of beer, or a bottle of spirits, remember that a forest had to be cut down to grow the grapes, barley, or hops that you're enjoying. France cut its oak forests to grow grapes. Germany, Italy, California, Canada, Australia, Chile - and the list goes on - all cut down their forests to grow grapes. We all have to look at how we use the surface of the earth."

John stood up and came to centre stage and thanked Jack for his presentation. "On the way out please pick up the list of essay topics and take your time deciding what you would like to do. Pick a topic and hand in your outline in two weeks. I've asked Jack to come back next Saturday to do a north-south profile. See you this week at our regular time and place."

Friday Hope was lingering for a chance to talk to Jack. Jack caught his eye. Many students were snapping pictures of the art work across the front of the room.

First Term Assignment:

General Science 135 sections A, B, C Professor John Henry Ph.D.

First Term Assignment: 20 %

All calculations please use both metric and British measures.

All assignments, or with pre-discussion authorization, must be kept to a maximum of five pages (be succinct)

- Calculate the green vegetation water sink (GVWS) in (pick a province or territory). Compare the historical GVWS to today.

- Calculate the water turns from west to east across this country. Speculate on the historical change in turns, providing your assumptions and background data.

- Design and describe a porous parking lot that will allow trees to shade the complete lot and water to penetrate the ground.

- Calculate the land area lost to highways, streets, and roads. Estimate the lost GVWS. (Pick a county, province, or township.)

- If buildings were clad with wood or any other water-absorbing material, estimate the amount of water that would be stored by not using glass, steel, aluminum, stone, brick, or plastic siding and roofing for commercial and residential buildings.

- Create a plan to promote the planting of our forests on private land.

- Formulate a plan to build back the planting of our forests on government land. Include in your plan the roadsides.

- Free grazing of animals will no longer be tolerated. Build a case to justify pen-feeding animals in shaded paddocks.

- What are the new measures to allow harvesting of forest products and how would you implement these concepts?

- How can we justify having a lawn or allowing golf courses to exist?

Optional:

- Discuss a topic with me that is not on the list but that you would like to tackle.

Please note all assignments must be done in groups. The groups will be a minimum of three students and a maximum of four. A group mark will be given and your group will determine how the marks will be divided among the members.

Each student will be required to sit on a panel and all presentations are open to members of the class. Student panel members are responsible for critical comments only. Grades are the sole responsibility of Dr. Henry, according to the rules and appeal procedures of the university.

CHAPTER THREE

Jack's Second Lecture

The following Friday night Jack was back sketching on the blackboards and John was in his office working on an article he hoped to get published. On the top of John's filing cabinet a pile of papers and a few USB sticks were waiting for his newly hired assistant. Many graduate students were eager to earn a few dollars to make ends meet. The funds were in the departmental budget.

The Icebox Theory of Global Warming

Jack was busy practicing and spacing his drawings and checking his notes.

Jack's notes were mainly pictures he had sketched at home in Richmond Hill. Between checking the ice trays in the back shop and troubleshooting calls he sketched his next lecture. Jack pondered how he should introduce the Icebox theory. It is so simple, Jack thought, if the audience can visualize the earth as a round ball floating in space. On two ends of the ball we have frozen water because of their locations. On top of the high mountains around the world there is frozen water due to the altitude above sea level. When he was finished his

103

board work for the evening, out came the Sunlight bar soap and the pail of water.

All set, Jack and John would meet Helen, John's wife, for beer and wings. Helen was eager to join them at the pub and chip in on the conversation. She did not like wings but could find a salad among the bar food. She found Jack refreshing and she needed the adult break away from the children. Jack had to tell them about Deloris's (D) new machine. D was the recent hire and she and John Henry were long-time friends.

Saturday morning Lloyd Richardson opened the door to the lecture hall. Lloyd had remembered to move the lectern and the stage was empty.

Once the lights were turned up, Jack started to draw from his notes. The sun, earth, moon, and local planets were all drawn in on the left board during the summer tilt of the earth's axis. Moving to the right, the fall, winter, then the spring position were in place.

Along the top of the board Jack drew in two glaciers moving down a mountainside, colliding in the centre half-way down the mountain and the two ice caps, north and south poles. Greenland was in one corner with its ice cap.

At nine-thirty, students started to arrive, filling the front seats. John looked up from his manuscript. The laptop was balanced on his knees. He was not used to students arriving early, let alone sitting in the front seats.

At nine forty-five the lecture hall was full and students were finding seating in the aisles and standing at the back. At ten the hall was packed and there was no room for anyone to stand. John looked up from his papers and saw his department head and three of his colleagues sitting in the centre of the lecture theatre. This was a first.

John walked to the centre of the stage and the audience grew silent.

"We have a full house. Thanks for coming out this morning. I hope Friday night was not too hard on you."

A ripple of laughter went through the audience.

John, "I would like to acknowledge our department head, Dr. Hendren, and Professor Cline from our economics department." The two professors nodded back and were inwardly pleased that they had been properly acknowledged. The other two professors knew that John would not have been able to recognize them as they had not met previously.

The other two professors were Dr. Janice Lloyd and Dr. Herbert Berger, both from the geography department. Bernie and Ralph, the two geography majors, had mentioned trippage, slippage and sippage in their last seminar class. The concept of trees being the lost water sink and the concept of green vegetation water sinks had piqued the curiosity of Dr. Lloyd. She had convinced her friend Herbert to come and pay a drop-in visit. They had decided to see Jack Wilson for themselves. Bernie had mentioned to Professor Henry that two of his professors might be dropping in to catch the lecture if they could free themselves from a department meeting.

❦

Jack said to himself, "Jack Wilson is back in town and looking forward to explaining his theory on global warming."

If anyone thought Jack was nervous the first time speaking in the lecture hall, well, this was pressure.

Jack said to himself, "These visiting professors are just like rink managers dropping in to get the local information on the new waters I'm selling." Jack parked his fear of failure at that moment.

"Please welcome Jack."

There was a thunderous applause and John could see it surprised the drop-in professors.

Jack touched his microphone and checked to make sure he had all his tools. A nod from the back of the lecture hall and Jack thought he was ready to go. Lots of chalk and a clean brush and I better get going. Joy, his wife, suggested that he wear light-coloured clothing when handling chalk as when he came home the week before, his clothes looked like he had rolled down the chalk cliffs of Dover.

Jack walked briskly to the front and centre of the raised platform, big fingers surrounding the large yellow chalk.

He had spent a couple of sleepless nights trying to work out how he would introduce his next lecture. Finally he came to the conclusion that time was precious and the students were there to learn. The least time spent on the introduction would be the best. The important thing was to get to the task and make the presentation interesting.

Jack paused and walked to the left of the stage. He just walked to the board and all eyes followed.

With his back to the audience, chalk in hand, Jack started to draw. "We are in summer and you can see the direct rays of the sun are hitting us, southern Ontario, in a very direct fashion."

The yellow chalk was drawing in the lines, and arrows were appropriately placed. "This is the science class you had in grade six. Let's take a few minutes to review the basic concepts of rotation, tilt and the effects of gravity. In a few minutes I'll assume you understand these basic principles."

Fall, winter, and then spring were quickly shown and the equinoxes and summer and winter solstices were noted on the board. Jack pointed out the two poles, emphasizing the total darkness in the winter and total sunlight in the summer.

Jack took a new brush he had bought. It was almost twenty-four inches long and Jack, with two walks across the front of the room, erased most of the blackboards. He started with his chalk in the top right corner of the board at the far right of the room and curved it down to the bottom left of the far left board.

Jack, "This is the arch of earth from north to the equator. You may have to slant your heads to visualize that this is north to south. Students from down under, you may want to crawl under your seat."

Laughter erupted from the New Zealand and Aussie students.

July was printed at the top of the board, "JULY"

Jack started to draw in the ice cap on the north pole and clearly indicate the ice on land and ice on water. Greenland and Iceland were drawn to scale and their ice caps were sketched as Jack explained how yearly, and centuries of, accumulate snow and gravity built up the ice sheets.

The chalk showed how the layers built up and the weight started to spread out the mass of white.

Jack, "Weight and friction pushing down created a wet slip layer."

This layer was shown in blue.

"This allowed the heavy ice on top to spread out. The warm rock underneath also helped form a slip layer of water. Arctic ice did not flow north, it pushed south. Antarctic ice flows north not south.

Jack turned to his audience and said, "Do you know why the ice in the north flows south and the ice in the south flows north? Why does the arctic ice not flow north as the north receives very little snow accumulation? As well, the south pole is a dry desert."

As the right side of the brains were listening, grasping for the answer the left sides were focusing on the evolving sketches on the board.

Arctic tundra was captured; the tree line, the forest line, and the agriculture line were laid out on the blackboard. Inches of topsoil in the north to feet of topsoil in the south were captured by the drawings. The land between, the strip of land that moves from the Canadian granite shield, to limestone, to the great lakes lowland was sketched on the board. It stretches like a hockey stick across the province of Ontario.

Jack was working fast and the boards were filling up.

The side or cross-section views of rivers and lakes and the great lakes were sketched into their corresponding relationships to the climate zones.

As their right brains were still pondering that question, the left brains were watching as Jack drew in the tropical sun, beating down on the rain forests at the equator. Clouds were rising and heading north. Cool winds were being drawn, moving down from the north and heading south. Warm layers of clouds were moving over cool layers coming south, and hurricane alley was in full swing. Rain and hail were dropping out of the sky onto the earth below.

Jack stopped then drew a glass of water. In the glass he put in one ice cube, in the identical glass beside it he put in two ice cubes. On the next board he sketched in a block of ice three feet wide and beside it a block one foot wide.

On the next board he drew two blocks of ice the same size, only one was twice as thick as the other one.

The Icebox Theory

Jack then turned to the audience and said, "This is the basis of my Ice Box theory on global warming."

Pointing to the board work, Jack continued "I guess I could have called it the bar theory. When you are out at the bar drinking and you have one cube of ice in your drink or two cubes, which one stays cooler longer?

When a wind blows over a block of ice three feet wide or one foot wide which mass of air would be cooler?"

"If the wind blows over a block of ice six feet thick or eighty feet thick, will there be a difference in air temperature between the two?"

Jack paused and waited.

"My theory is very simple. The sun is melting the polar ice caps and the sun is reducing the surface area of the ice sheets. The cool winds from the north are not as cool because the ice sheets are smaller. This is just one and only one contributor to global warming. When I was a young boy helping our family make a living, we cut ice in the winter and delivered it in the summer to our customers. If you wanted to cool a large quantity of food, you had a big ice box and you bought a large block of ice every two days."

Jack was quickly drawing two ice boxes side by side. "Smaller ice boxes with less food required a smaller block of ice. How many days a year is the temperature above freezing at the poles? How many hours? If you increase the air temperature one or two degrees how much more ice can the warmer air melt? For 335 days a year the temperature is below freezing. If the temperature warms up from -23 to -22, nothing is melting.

It's not the air temperature; it's the rays of the sun. Frozen water on the mountain tops is being burned off by the sun. Have you ever watched in the spring when the winter is retreating, how the sun comes out and cooks the ice and snow banks? Snow and ice go from a solid to water vapour, skipping the water stage. Sublimation is reducing the size of Earth's natural ice box. "Evapotranspiration" is the word used today. Once the sun's rays reduce the size of the ice block, we're back to the large ice box versus the smaller ice box. The smaller ice box can't do the same job and guess what, it's warmer down here." Jack points to the smaller ice box on the board.

Jack sketches the hot, wet air rising in the equator and the winds blowing the clouds east.

"On our west coast the glaciers are formed by layers of snow over time. There is no body of water to freeze at the mountain tops.

Do the recorded statistics show we are getting less snow? Do the records show the change in the number of hours of sunlight over the years for a calculable comparison?

Is it warmer on top of a mountain? The day I did a face plant at Whistler the wind-chill factor was minus fifty-two. Minus fifty would not have felt any warmer or melted any ice or snow. It is the sun that is eating the snow and ice."

Jack went on to explain how he played with a number of solutions to block the sun rays. He explained with diagrams how he was able to hold sections of the snow banks along the side of the road two weeks longer in the spring. How a ski resort was able to protect an exposed side bank and have the snow stay longer during the spring break.

"You have seen snow banks on a country road. On a country road running north and south, have you noticed in the spring which bank starts to melt and disappear first?

Block the sun and Earth will hold the ice and snow. Is this a provincial or a federal issue?"

The class laughed and applauded.

"This is called dodge ball. When you want to stall or do nothing, bring up the old easy way out. Question whether it is a provincial or federal matter long enough and the issue will go away.

Remember we were talking a couple of weeks ago about the water cycle turn over. If you correlate the reduction in water turns, or trippage, to the earth's temperature, you might find some interesting numbers."

As he printed on the board, Jack continued. "In the last lecture I talked about shade, evaporation, altitude (SEA) as being the only way the planet can cool itself. How does this cooling theory apply to the polar ice caps and ice-covered mountains in the world?"

Jack printed Shade on the board. "The rotation of our planet on its axis once every twenty-four hours means that every spot on earth over the length of one orbit around the sun gets the same number of hours of sunlight and the same number of hours of darkness or shade. When the north end of our planet tips towards the sun we call this summer and the far north experiences up to twenty-four hours of sunlight. The far north has very few frost-free days. To think or believe that the air temperature is melting the caps is a big stretch. The sun's rays are coming in at quite a slant and they're not as hot as the direct rays nearer the equator. Guys and girls, you know all about this. If you walk into the pub and you get a glance or a direct stare, you know which one is the hot one."

The audience broke out in laughter.

"The Ice Box theory is very simple. The sun is reducing the size of the ice blocks around the world and the box (planet earth) is heating up. We have a choice. We can block the sun and keep the ice. We have the option of doing nothing and living through the warming cycle. Ladies and gentlemen, it's your call."

Jack then spoke to the geography students. "In the far north we've discovered oil and gas and it seems like there is plenty of it. So much so, that the northern countries are going to be arguing over who owns the Arctic. At least seven countries are claiming part of the Antarctic for the same reasons. If you believe the theory that rotting vegetation or rotting animal material is the basic building block for creating fossil fuels, I have questions. In the north, with no light and cold how did anything grow? If the fossil fuels were created in the south, how long did it take our continental plates to drift north and still have enough time left to create fossil fuels in the south? The numbers do not add up for me. Do they add up for you?"

Our Forests

Jack was listing some of the functions of our world forests. On the centre board Jack printed:

WORLD FORESTS (80% DESTROYED BY HUMANS)

1. Carbon dioxide to oxygen. Air purification.

2. Ground water is sucked up and evaporates. Water purification, cooling, and trippage increase.

3. Trees provide shade. Cooling.

4. Trees are green water sinks, holding forty-five to eighty percent of their weight in water. Trees are also carbon sinks. The world's largest sink of fresh water was in its tree cover. We have skinned planet Earth.

5. Forest floors hold water and hold our top soil in place.

6. Excess water seeps down and fills up the ground water table.

7. Trees are the world's air conditioner. (80 % removed).

Jack sipped on his glass of water and then turned to the board.

"As the water evaporates from the trees the vapour moves upward. The higher altitude cools the water vapour and eventually the vapour coalesces with dirt and dust particles. The cool rains fall. Rain and snow not only clean, they lower the temperature of the earth's surface. The evaporation of the surface rain water further adds to cooling our planet. Now you know why, when you collect rainwater, it is a brownish colour. This is one of earth's natural ways to cool down and clean the air we breathe. Why would we want to prevent evaporation? Should we let China know?"

Large laughter erupted from the audience and John smiled at Jack.

Jack held up his hands like a preacher on Sunday morning in church. "Water comes and it goes." It was not a lecture, or a sermon, it was a statement of fact.

Jack looked into the audience and slowly said, "Remember all the fresh water in the world comes from the oceans and seas. Gravity pulls all the fresh water in the world back to the oceans and seas. This is the big water cycle. The rest are subsets of this world cycle. The trippage of those water molecules is what we are destroying when we cut our forests. We're preventing the planet from cooling itself.

Just to refresh your memory, many of you have walked in the woods during the winter and watched as the snow hangs onto the needles and limbs of trees. Eventually the snow and ice falls to the ground. In the spring that snow and ice melts. A large percentage of the water soaks in and replenishes the groundwater table.

Snow that falls on a farmer's fields is often evapotranspired by the winter and spring sun. On really hot days it melts quickly and runs off the frozen ground into streams and away it goes. Our land use practices increase sublimation and reduce melting and groundwater. The snow in the forest beside an open field melts slower. The air temperature-not sun's rays-melt the snow and ice in the forests. This is the big difference. The water is released and more of it hangs around. The forests replace the groundwater. Farm fields reduce the groundwater. We pay a price for farming in open fields.

In the summer, why is it cooler in the forest than walking across your lawn or parking lot?

The sun is hitting the leaves above and working its magic, photosynthesis, to grow the trees. Leaves allow moisture to evaporate. You are on the forest floor and the moisture from the trees and ground is evaporating. This evaporation process is like your body sweat, evaporating and cooling your body down. We have sweat glands and we cool ourselves down while your pet dog is busy sticking his tongue out to cool down, as dogs only have sweat glands on their feet.

The water evaporating in the forest sucks the sun's heat away and you're cooled. The sun is not hitting you directly. You are not baking in a microwave oven.

Why do we have so much drinkable water? Thank the tree root system and the evaporation, cloud, altitude, rain cycle.

Why are we still cutting down our forests? We need forest products, but we need to understand more fully what we're doing. We need to manage our activities better.

❧

Just as an aside, when I see a beautiful lawn I think of the millions of dollars spent on advertising to sell lawn seed, fertilizer, lawn mowers, and the list goes on. We have been brain-washed into thinking a green lawn without weeds is what we should have. I tell my wife I'm going out to help the world, sit under our tall oak trees, and enjoy a cool glass of water. My neighbours are sitting on their riding lawnmowers destroying our planet. They're burning our fossil fuels, polluting our air, draining our water supply. Get off the lawnmower and get on your bike. I'm a big talker!

Inuit people had only animal skins to make their clothing. In the cold arctic their double-layered animal skin suits kept them warm and frostbite free. Skins are a renewable resource unlike petroleum-based clothing. This is a topic for another day."

Jack looked over at John and then the audience. "I have to stop and have a break. If you come back after lunch, I'll talk about a real problem we have and how you can help."

Lunchtime had come.

Jack wiped down the board. He grabbed a quick sandwich and a bag of Piccard's peanuts. Next to fresh cheddar cheese curds from Maple Dale cheese factory in Plainville, Piccard's peanuts were his ultimate treat. John was off to his office to return his phone messages and polish his manuscript.

On the board Jack printed in large capital letters.

SEPTEMBER: THE WORLD'S BIGGEST AIR CONDITIONER

The Canadian topography was left on the board while the weather patterns were erased.

Once the students were settled in their seats, Jack said," Here we are in Ottawa and fall is in the air. Cool weather and lots of rain showers.

My son Chip is taking evening courses at Atkinson College, York University, and his first year Geography teacher is a consultant to companies that are building weather forecasting models. This will not be news to our geography majors but Chip was surprised at the sophistication of the data in the models.

Hot air does not rise: Just put in the back of your mind that our weather patterns are due to three factors: temperature, planet rotation, and gravity. (TPG) Temperature is heat from the sun, and coolness from shade, evaporation, and altitude. (SEA)

Gravity just pulls everything around. Water vapour is lighter than normal air and the heavier air, due to gravity, pushes up the lighter air. If water vapour was heavier, we would have permanent wet surfaces and life as we know it would be completely different. When you take a hot or cold shower have you noticed that the mirror in the bathroom shows condensation? Condensation starts at the top. Then as you continue to run the shower, the condensation works its way down the mirror. When you step out of the shower the floor is dry. If water vapour was heavier than air, your bathroom, laundry room, kitchen, and any other room with evaporation would have wet floors. Water vapour condenses on the cooler surfaces so your cool glass mirror condenses vapour, but your wood cabinets do not.

Rotation of our planet allows our great ball to spin in that rich lubricant called the atmosphere. Have you ever wondered why most of our weather comes from the west. I don't have to put a spin on that story."

Jack continued talking while drawing the corn, soybean, wheat, oats, barley, potatoes, and other crops from west to east, across the country.

Trippage-don't cut your tallest trees

Jack caught his breath and began. "The weather forecast models track the activities on earth. Crops, the number of acres, and the time they are planted in the spring form large variables in the computer model. Pictures from satellites allow forecasters to determine the germination, growth, maturation and harvest times for various crops. Water absorption, water retention, and evaporation rates of each crop are programmed into the model. The crops are monitored and readings are taken daily. The first crop to be cut is hay in the June hay season. Some farms are able to cut three crops though most fields only produce two crops a year."

Hay, it is raining.

Jack, "You'll notice many farmers cut three or four rows around a hay field. The next day they bale the hay from those rows. Then in a couple of days they cut the rest of the field. This is a drying technique. Cutting the outside of the field first allows the outside ring to build up heat, then the centre cut will dry faster with the movement of air over the newly cut moist hay. The air flows from the centre out to the warmer edges of the fields. Farmers also get straight cuts down the field and turn on the outside rows, so there is less hay wasted on the swing corners. Jack hastily drew the hay fields while the farm kids in the audience were drifting off to their own farm fields. "Farmers don't realize that when they cut hay they are causing rain. Hay, which is a mixture of tall grasses, is over eighty percent water. Farmers cut thousands of acres or thousands of hectares. Where do you think all the water goes? The moisture from the hay evaporates and rises forming clouds. Given a chance, the clouds rise, cool, and then drop rain on the farmer downwind. Each farmer, following what they think is best, actually causes problems for their fellow farmers downwind. What would happen if they could get together and plan the hay harvest to not cause rain for their downwind neighbour? Every continent has prevailing winds. The cutting and drying cycle could be planned in such a way that the hay would be baled and stored and the rains from upwind would arrive to start the second growth. Do you think you could convince farmers to cut and harvest according to a schedule?

In the fall when the corn, soybean, and other crops mature and start to die, they shed their water moisture. Measure the square miles of crops that are maturing and shedding their moisture, add to that the deciduous trees that are starting to drop their leaves and the coniferous trees that are dropping their two year old needles. This huge water event is the largest air conditioner in the world. The sun and heat of the summer are required to turn this water into moisture vapours. This world-wide air conditioner cools down our planet. Our Indian summer only happens if the moisture is converted into vapour and the sun can get through the clouds of moisture before the end of October."

Jack became a guest speaker for John's classes whenever he was driving through town. At first Jack was quite shy and reserved in front of one hundred and fifty university students. Once question period opened up, Jack was in his element and soon realized the students were interested in what he had to say and sketch. He broke through the shyness and became very entertaining as well as informative; he sometimes left the students with more unanswered questions than they were comfortable with.

Jack: "John, how am I doing with these guest lectures?"

"Jack, you had the house in tears when you told them the story of the two pines in Buckhorn."

On the west end of Buckhorn on Lakehurst Road, there's a fire hall that services the community. The firefighters are paid volunteers and are paid to train as well. Jack was going by the hall when he noticed the fire-fighters trimming up two forty year-old pine trees. He thought, okay, they needed the air to move and the pines would continue to grow and hold water. The next day on his way home he drove by and there were only the stumps and a pile of sawdust.

Trained firefighters dedicated to saving lives were unaware that they had destroyed a forty year-old water sink that was holding conservatively one hundred gallons of water. Water so vital to their role in putting out house fires was not recognized when it was being stored in a tree. Jack knew his work was just starting.

The Marshmallow Example

When you are sitting around the campfire at the camp in summer roasting marshmallows, notice when you get the mallow too close to the fire it burns. Think of the marshmallow as the earth going around a hot sun. How do you toast the complete marshmallow and not just burn one side?

John, "Don't know if you've noticed but my students put the word out when you are coming. We have drop-ins that fill all the empty seats and sometimes there's just standing room only at the back. My department head asked me when you would be returning. That is as big a compliment as you are going to get. Jack a few of my colleagues in the science department want to know if you will do a guest lecture for their students."

Jack, "I have nothing new to say. Why don't you trade classes and give them the lecture?"

John: "Material and ideas are only part of the learning experience. Jack, students see your passion and commitment and they're attracted to this inner mission. Most people go through life without passion or love and are trying to find it. You have it in abundance and want to share.

Your satirical off-the-cuff comments lighten the burden of your thoughts. You include people in your dreams and you jar both sides of their brains. To be a good teacher you have to be knowledgeable, a comedian, and a salesperson. Your passion for what you believe in is infectious."

Jack, "They should meet my son Chip."

Jack, "Has your load lightened with the new graduate student you hired to help you mark essays?"

John, "Yes and no. I spot-check essays, some from the low mark pile and some from the middle and top, to see if there is consistency as I am ultimately responsible for the grades. My time is now spent trying to get a paper published. You know the old publish or perish rule is still in place today."

Jack, "What are you working on?"

John: "It is on water flow. Pipelines, water hoses and any vessel that allows water to flow has an effect on the water volume. I'm experimenting with the texture of the inside of fire department hoses to see if the volume of water can be increased by changing the inside surface. I've made some progress. Someday if you can make the time I'll take you into the lab and show you what I've come up with."

Jack was going to make time as he was working on another theory and this information just might help.

CHAPTER FOUR

Essays to Read

John met Jack at the local coffee shop. John had a black leather case full of first term papers; this time they were for Jack.

Jack, "Hope you don't mind meeting here."

Jack ordered two large coffees and a half-dozen assorted muffins.

John, "Normally my students just hand in a stick or e-mail their essays but this time I requested that they print a hard copy as well. I mentioned that you travel and a paper copy would be more convenient. Many of my students have asked if you'll comment on their essays. Jack, you've piqued their inquisitiveness. Every teacher hopes for this moment. It doesn't happen often. I hope you'll take the time to write them a note. Just take a pen and scribble something as they will read every word and phrase. My marker, Bob Little, graded the papers according to the university criteria but your comments are what they want. To a student they have all volunteered and signed up to take part and to critique with you the papers of their fellow students. I hope you have the time. There are twenty five group papers."

Friday Hope's essay was on the top of the pile. Jack noticed that John had not put any grades or comments on the paper copies.

Jack, "Reality sets in. How long do you think it will take per paper?"

John, "The good part is the papers are all short but one. The students limited their essays to five pages. The green parking lot paper has a number of attachments that I am sure you will find interesting. Most of the students said that condensing their thoughts down to five pages was the hardest part of the assignment. I think you can expect them to talk a lot when they have the open floor. If each of the group members talk to their part of the essay for five minutes, and the question and answer period lasts for 20 minutes, it will take a minimum of an hour per paper."

John: "My advice is to read the first paragraph where the students tell you what their essay objective is all about and then read the last page where they summarize and make conclusions. If your interest is piqued then you can go back and read the body for further information."

Jack: "Why not schedule two hours per paper? If we do four groups a day, three weekends would do it. I'm looking forward to the discussions."

John lifted the pile of papers and passed them to Jack.

Jack would read every essay, word for word. Initially he was reluctant to make comments directly on the essay, but soon he was using and filling the margins and back of the sheets to extend his comments and thoughts. Many times a diagram was put on the blank back of a page.

⟨❧⟩

The next day when the shop was running smoothly, Jack reached for his folder and pulled out the first essay.

⟨❧⟩

General Science 135 section (A) Professor John Henry Ph.D.

How to Implement Green Vegetation Water Sinks (GVWS)

By:

Andrus, Anson

Bouskill, Audrey

Jaworski, Leonardo

Lazure, Monique

Jack was looking forward to reading this essay but the phone rang and Chip was asking for help. A winter storm had grounded all the flights out of the airport in Calgary. Chip was not going to be able to make it to the next job in Chicago. Jack checked the schedules with Sara, the office manager. They soon had a plan to shift Mat, who was a few miles away from Chicago, and Jack would pick up Mat's next job in Sarnia. Sara would phone Mat's wife and Sara would end up driving their oldest daughter to her lessons while mom looked after the younger tykes' hockey schedule. Chip would wait out the storm and catch the first flight home. Winter conditions often interrupted the best laid plans.

Jack quickly went down the list of materials he needed to take and contacts he required. Sara couldn't book a short flight to Sarnia so she booked Jack on the train out and the bus back. Taxis would bridge the gap. Jack figured if he could stop talking to people on the bus and train, he could sit back and read the essays.

Later that day Jack was packed and settling in to his seat on the train. The coach car was about half full and he had the temporary luxury of spreading out over two seats. He guessed that on the next stops the empty seats would be filled.

He opened up the first essay and started to read.

Introduction:

Logging companies today clear-cut or selectively cut based on sustainable supply. Renewable forests are required to supply building materials and products for a diverse economy.

Starting today maintaining and increasing GVWS must be factored into forest management plan.

How do we implement GVWS (Green Vegetation Water Sinks)?

Jack flipped to the back page and read:

Conclusion:

Tree markers must be educated on how to cruise a woodlot and determine the volume of contained water, the retention rate, and the annual accumulation. Harvests must not reduce the contained volume of water but only extract less than the accumulated annual increase. Mature forests at one hundred per cent GVWS may take out only the projected annual growth of accumulated moisture.

Provinces and territories are all responsible for their own forest and forest industry. The Ontario forest department was gutted by a former premier. Only a skeleton remains and the provincial government relies on associations like the Ontario Woodlot Association to carry the responsibility to maintain private forests in the province.

Jack then turned to the front page and read the complete report. He would do the same with all the other reports. He would read the opening, then closing, and then read the complete report. Jack was enthralled with the thinking in the body of the essays. Jack thought to himself...four more people understand GVWS and only eight billion more to go. You have to start to finish.

༄

General Science 135 section (A) Professor John Henry Ph.D.

Brock, Roland

Foster, Joyce

Porter, Neil

Wheeler, Howard

Introduction: The Plan to Re-Forest Southern Ontario

We quickly realized that each region of each province, state, or country will have to work with the resources they have and the tree seeds they can propagate. Ideas, solutions, and methodologies might work in more than one area. We hope the free flow of communication among the various bodies and institutions will benefit everyone for the common goal, putting back the forests.

Conclusion:

I. One organization in each country should be set up to be the communication manager among all their own groups. Thousands of concerned and dedicated citizens will be required to gather the appropriate healthy seeds, and grow the seedlings. Fast, efficient distribution of the seedlings to the planters, coupled with the organization of the actual planting, has to be carefully planned.

Check List

This is a simple project with a simple move - Put Them Back. Planting trees is done every year by squirrels, chipmunks, and other food-storing animals, even rats and mice. This is not a complicated exercise but it is time-sensitive and will take years to replace the trees which are gone. Involve all the different groups on the planning committee and hold open meetings. Seek expert advice wherever and whenever you can get it.

Besides the squirrels and chipmunks there are many tree experts in everyone's community. Loggers, sawyers, nursery owners, farmers, arborists, and the list is long. In the past, the squirrels and chipmunks dug holes and stored away their winter supply of nuts. Due to poor memory they've been the major tree planters.

Form a management group with a clear objective:

Increase the GVWS in your county, village, town, or city.

Landowners, students, public representatives and private landowners (majority of the members must be private landowners.)

Committee # 1-Determine the scope of the project and set realistic achievable goals for the village, town, city, or county. Be prepared. This is a minimum five year, maximum fifteen year project to get the areas planted that need to be replanted. Remember, membership will fluctuate as students leave for college, jobs, and university. Older members will retire.

Committee # 2-Publicity of the project to community members. Identify the large water storage trees existing in the community. Start the "Rain Drop Program". Where possible, every homeowner will want to put a rain drop on their largest water sink.

Committee #3- Gather healthy seeds and grow seedlings. People willing to grow five or ten seeds are just as welcome as landowners that wish to grow thousands. Determine what trees grow well in your area and in what soil and water conditions. Grow the same variety from locally gathered seeds. Local seeds will do best. Trees that grow well in an area have adjusted and the hardiest have survived. Introduced seeds from other microclimates or zones generally don't do as well.

Committee #4- Planters could be grade six students and retired community members. Spring coniferous planters and fall deciduous planters. Identify the areas to be planted first and set up the schedules.

Acquire sponsors to help defray some of the costs: seeds, fertilizer, and transportation of the seedlings, planters, and equipment.

Organize the digging of the seedlings, packaging, and shipping.

Organize the planting days.

First-year and second-year monitoring and maintenance.

Yearly pruning, culling, and replanting.

Identify the lands that need to be replanted.

Identify the roadways and acquire municipal and provincial/federal approvals.

The Ontario government tree seed facility at Angus does not have enough seeds or room to plant all the seeds that are needed for the province. They only have a billion seeds in inventory, less than one year's supply for the province.

Rain Drop Program

Any community, service club or family can adopt the Rain Drop Program. Every property owner or renter can participate in the awareness program. The rain drop will be mailed out so it can be placed on the largest water sink (tree). The rain drops can be made from non-rusting material and on the back the date and volume of water can be recorded. The rain drops are designed to last a very long time so the rain drops can be moved from tree to tree. Screws or nails should not be used to place or hang the rain drop. A ribbon or cord will prevent future problems when a tree needs to be cut down. The future sawyer or arborist will be thankful. When trees are taken down... splash not timber is the call.

〜

How to justify golf courses?

Conclusion:

One hundred to two hundred acres of farmland or low level stream land turned into an ornate back yard for someone to hit a ball can no longer be justified.

The planet has seven billion people and one quarter of them are starving. The land that is presently used by golfers could be growing food or trees. This could result in lower ocean and sea levels. The fertilizer, water, and machinery that are consumed for recreational purposes cannot be justified.

Recommendation:

Until all the people in the world have been fed and cared for all land used for recreational purposes should be banned. This includes not only golf courses but also soccer, football, baseball, cricket, polo, and lacrosse fields.

Only after everyone is fed should the United Nations allow land to be used for recreational purposes.

We recommend a ban on all professional sports until every last human on the planet is fed every day. Shifting the billions poured into baseball, basketball, soccer, football, tennis, golf, and hockey would soon create an urgency to find a way to put in place a distribution system so every person could go to bed every night without hunger pains. The world produces more than enough food to feed every person on the planet. More food is wasted and rots than is required to feed the hungry. Feeding the hungry is not in the best financial interests of large food companies. Companies are run for shareholder value, not for what is best for the people of the country. Political parties, once elected, will protect their financial sources for the next election. Governments will often appoint heads and retired executives of these food companies to sit on important government committees. This is much like the fox in charge of security for the hen house.

Jack would spend a couple of days making comments on this essay. Idealistic students can move mountains and this would be a powerful rich mountain to surmount. Sports teams around the world are owned by the wealthiest people and the only thing they understand is money.

Jack read the balance of the papers and was impressed by the creativity and logic of the young adults.

One conclusion caught his attention:

Green Vegetation Water Sink (GVWS) and Forest Management

Conclusion:

Each year you can only harvest less than the incremental increase in your total GVWS in the forests that are presently in place. For forests that are at their maximum GVWS you can only harvest the annual growth rate of the forest. If your forest will take sixty years to replace, you can only harvest one sixtieth. Likewise, if the mature forest is one hundred or two hundred years old to replace you would be able to harvest one hundredth or one two hundredth per year.

Jack found it both encouraging and fascinating to read about the students' varied approaches. When he came across the world's largest lawsuit he smiled...

The world's largest lawsuit. (Statement of Claim)

Trial to be heard at Le Hague, The International Court of Justice.

The Plaintiff: All the islands and countries flooded and threatened by further reduction in the GVWS; joined by Holland, Italy, and at the last minute, India.

The Accused: Europe, Russia, China, Canada, United States, Asia, South America.

Claim # I: Cutting down seventy five percent of the worlds' trees, thus reducing the freshwater supply of the world and increasing ocean levels.

Claim # 2: Reducing the natural air conditioners and thus responsible for heating up the world.

Claim # 3: Cutting trees and as a result reducing the natural carbon storage capacity of the world.

Seeking: Stop action and provide compensation for the damages due to flooding.

Special Note:

Our group feels there might be an ulterior motive for the industrialized countries to continue flooding out the small island states. There is much to be gained by resource-based companies to have free access to fossil oil, minerals, and metals when these small islands disappear and then become international zones. We quote Wikipedia.

Wikipedia Quote: Besides the issues that flooding brings (soil salinization, …) for these islands states, the islands states themselves would also become dissolved over time, as the islands become uninhabitable or completely submerged by the sea. Once this happens, all rights on the surrounding area (sea) are removed. This area can be huge, as rights extend to a radius of 224 nautical miles (414 km) around the entire island state. Any resources (fossil oil, minerals, metals, …) within this area can be freely dug up by anyone and sold without needing to pay any commission to the (now dissolved) island state.[104]

http://en.wikipedia.org/wiki/Current_sea_level_ rise#cite_note-104

The planet must have a warm core or the planet is dead. Our group concluded that the pumping out of oil and natural gas warmed up the pockets of many world giants. Removing the crude oil and gas which is under high temperatures and pressures is cooling down the core of planet Earth and on the surface billions of barrels are helping heat our planet even before it is refined and burnt.

Jack thought to himself, the invisible hand is at work again.

CHAPTER FIVE

Off to Ottawa Grade Sixers

John Henry's friend, Ricarda, was an elementary school teacher in Ottawa. She and her best friend Barbara were to pick up John for their pre-planned luncheon. They were informed by John's marking assistant, Bob Little, that they could find John in the lecture hall. Ricarda and Barbara slipped into the lecture hall and were watching the students, using their phones to take pictures of the drawings on the black boards. As Ricarda was listening to Barbara and John get caught up from last term she couldn't help but notice the diagrams on all the blackboards.

Ricarda, "John, oh, it looks like you've been drawing."

John, "Absolutely not me, you know; I can't draw a realistic stick man."

Ricarda; "Who did this work?"

John, "My friend Jack. When he's in town he's our favourite guest lecturer. He just left a few minutes ago or I could have introduced you both to him. You would find him very interesting. Before you ask, he's too old for you, and he's also happily married."

Ricarda, "Next month Barbara and her best friend Julie are taking their grade six classes to Ottawa for their annual field trip. This year they are both teaching grade six in Norwood. Barbara teaches at the public school and Julie teaches at the Catholic elementary school. They will combine their classes to fill the bus and reduce costs. They're arriving on Friday and staying overnight and leaving late Saturday."

Ricarda, "Would your friend talk to grade six students?"

Barbara, "We want to bring them here for your tour and maybe experience a different learning style. Many of these children have never been to visit a university. Some have no one in the family that has gone to college or university, let alone graduated from high school. We have a number of students this year that are struggling with language arts. They would connect with these diagrams."

Barbara; "Don't get me wrong. This crop of grade sixers is a pleasure to teach. For some reason, since kindergarten, they have just been an enjoyable group of children. It's fun to go to work when you have a class like this, unlike the grade sevens ahead of them. Every teacher gritted their teeth when it was their turn to try and live with them for a year. There were two bullies in the class that dominated all the other students and the games that were played by those twin girls had the class upset most of the time. The boys were acting out and they didn't even know why. As you know, the boys are often socially one or two years behind girls in elementary school and can be easily out-manoeuvred by a couple of female bullies."

John, "It sounds like being married." The three of them laughed. "I'll ask Jack and I'm sure if it's in his power he would be very willing. Jack travels all over the country painting on ice.

What topic or topics would you like him to talk about? He would be able to speak and draw on the board and tell them a few stories. You know Jack grew up in Norwood. He is not a teacher or a professor; he owns and operates an ice painting company in Richmond Hill. Jack has learned over the years to teach and motivate his employees. He worked for his neighbour mixing paint and one thing led to another. After schooling was over, Jack worked full time for his neighbour, then eventually took over the company when his neighbour wanted to retire. He moved the operations out of the garage into an industrial mall in Richmond Hill to be more central to their ice painting market."

Ricarda, "What's his last name?"

John, "Wilson."

Barbara, "There are lots of Wilsons in the Norwood area. One of my all time favourite students, Dougiee, is more or less being raised by a really old man called Mr. Bob and he's related somehow to the Wilsons. You will like Dougiee. He's everyone's favourite as he is just Dougiee. Teachers are supposed to not have favourites, but you'll meet Dougiee and you'll see. Mr. Bob and Grace rubbed off on Dougiee."

John; "You know, Ricarda filled me in about Dougiee. Fire fighters, when they get together, relive all the past fires and teachers talk about students, the good and the bad."

❧

In a previous conversation.

Ricarda, "Douglas Victor MacArthur is the name his parents gave him when he was born. The parents, both professionals, are very involved in their own careers. Vera, his mother, is on a career path to be chief financial officer of a large expanding land developer. Dr. Victor MacArthur, his father, is completely absorbed in his medical practice in Peterborough. He's a heart specialist and has the attitude to go with it. He is tolerated because he is one of the best in his field. Patients refer to his bedside manner as being like an ill-mannered pit bull that has been crossed with a jack ass. Dougiee was slow off the start line and did not meet the high expectations of either parent. He was ignored, almost from birth. They hired a local widow, Grace, who needed a job and the money to day-sit Dougiee and basically raise him. They would leave in the morning before he was up and come home in the evening after he was put to bed. Each day, seven days a week, Grace would pick Dougiee up out of his crib and take him home for the day. I'm not sure the parents ever changed a diaper.

Saturdays they worked and Sunday after Sunday school would be the only time Dougiee would see his parents; it was a scheduled fifteen to twenty minute session. Sunday dinners were important business and social times. Dougiee would often be introduced, and then excused, and he would go to Grace's or Mr. Bob's for dinner. Grace's cousin on her mother's side was Reg Wilson and Reg soon had Mr. Bob in the picture. In small towns, you know, just about everyone is related to everyone. Reg Wilson is Jack Wilson's dad. Mr. Bob is Reg's best friend and first cousin on his dad's side of the family. Mr. Bob likes kids and needed a new challenge. It all seems to work. You've heard the expression that it takes a village to raise a child? Mr. Bob and Grace are the core of the village for Dougiee."

In a small village just about everyone is related to everyone else if people have the memories to go back. New undertakers in town soon find out who knows who and for the first few years they are called upon many times to make the necessary connections. Mr. Bob was called on many times to sort out the family tree for Kelly, the new funeral director.

Mr. Bob was from a large Irish family and he had seen neglected children, but they were usually in families with ten or more children. Parents could not become attached to babies or children as so many died before they became teenagers. The pain of losing small ones was too much. Parents did not bond with their children until they felt that they were here to stay and work. In large families the older siblings were in charge of rearing the younger ones. The older brothers and sisters became attached to their younger brothers and sisters and provided the caretaking and the nurturing. Mr. Bob and Grace stepped in to be parents for Dougiee until, or if ever, the parents connected with their only son.

Mr. Bob saw Dougiee every day and just about all day, when he was four and five years old. This was before school would steal him away. Dougiee spent most of his day following Mr. Bob around on his farm. The farm was just at the end of the street. Mr. Bob taught Dougiee how a farm operated and how to care for all the animals. Dougiee had his own chicken, a young bird that grew from a yellow ball of fluff into a large egg laying machine, as well as his own rabbit, and other animals. Dougiee raised a short-horn heifer and showed her at all the fairs. Queen nearly made it to the Royal Agricultural Fair in Toronto.

Barbara; "Dougiee is by far the brightest student I've had the pleasure to teach. It just so happens that this year I also have Allen and he can keep up to Dougiee but he is so quiet you would never suspect his ability. Both of their IQ test results are just above average because the tests ask all the questions Dougiee and Allen don't know. If you want to know how many feet in a rod how many pecks in a bushel how to graft a new apple branch onto an existing tree, and the concession road is a chain wide then the score would be off the chart. Dougiee and Allen never miss a math or arithmetic question. All tests are language and culturally biased."

Mr. Bob sort of unofficially adopted Dougiee. You can tell Dougiee has wormed his way into many lives but not his biological parents.

∽

Phone calls were made and schedules changed, and Jack would be willing to try to entertain and maybe teach the grade six classes. Jack wanted a list of all the students and staff who would be attending as he was curious when he heard they were from Norwood, his home-town. Maybe he would know their parents or relatives.

Jack remembered a conversation with his son Chip that Hank was working at the Norwood curling rink. After retiring, Hank took on the new night shift manager position. Jack thought Hank may have taught grade six or grade seven. Hank had taught Jack in grade seven a long time ago. Jack was thirteen years old and his very old teacher was Hank. Hank had just turned twenty and was in front of his first class.

Jack dropped in and sure enough Hank was opening up his thermos and pouring a coffee.

Jack, "Hank, what do you spike your coffee with?"

Hank, "Spike? Coffee, grab a cup."

Hank drank the strongest espresso, black. It would keep your eyes open for a week. Hank chased the espresso with hot water during the day, and at night, when he was at the end of his shift, with warm rum for the cold lonely walk home. Hank had lost his best friend, Jack Payne, when they were teenagers. Hank did not want to get close to anyone else again and picked a life with people, but inside himself. It may not be a good example but many people after losing their favourite pet do not want to go through the hurt again so deprive themselves of the joy they once had. Hank also decided not have any pets in his life.

Jack took a sip and said, "You haven't lost your touch."

Hank, "Drank too much staff room swill; you need a bit of bite."

Jack handed Hank a paper bag with two pounds of green coffee beans.

Hank looked into the bag. "This is the good stuff. Thanks. Why are you here? Your Rats painted the ice last week."

Jack, "I need your help, I've been asked to teach a group of grade sixers. Didn't you teach grade six?"

Hank, "Taught grades six, seven, and eight, for thirty-five years. Only met one student that I couldn't teach."

Jack, "Who was that?"

Hank, "You."

They both laughed. Jack prided himself in the ability to tease everyone he met, but Hank always was one step ahead of him.

Hank graduated from Peterborough Teachers' College. This was before the universities muscled in and took over teacher education for elementary teachers.

Hank, "Keep it simple, don't repeat yourself, keep moving, and if you lose someone, circle back and catch them up.

IPSA Peterborough Teachers' College: Introduction, Presentation, Summary, Application.

Keep your introduction short but catch their attention. Present your material in three ways: visual, auditory, and action. Summarize the key concepts at the end but don't repeat your presentation. Do it differently. Application is most important. Find a way to apply your knowledge. Homework is a killer; make it interesting or don't give any. Give them a job. That age group likes to work."

Jack; "I have about three hundred university students presently doing research for me. Maybe I can come up with something for these young students."

Hank; Use the motto, "We learn by doing". Get them out of their seats as much as possible. If you're teaching sixty students in a lecture hall, remember that combined they will know much more than you, and all you have to do is to draw them out."

Jack; "Any more hints?"

Hank; "To be a good teacher you only need to remember three things: you must be a clown, a salesman, and an actor. You can't fool children; they will get to know you. You're good with names. Read over the list of students and be able to pronounce all their names, first and last. I made a practice of memorizing the names of all my students the first class each September. In a small elementary school I knew twenty-seven out of thirty-five names before they entered my room.

Hank, "What topics are you going to address in your lecture? How to paint ice?"

Jack, "The two topics that are on the curriculum and not yet covered by the grade six classes are: the body cooling system in health education, and crop rotation in the science/agricultural experimental course of studies. I can take my pick."

Hank, "Cool down the world. See if you can motivate those kids to plant some trees. You hook them at that age and you have them for life."

Jack and Hank caught up on all the local news: deaths, births, trips, and gossip. Morning would soon be arriving. A warm toddy off the hot water boiler in the back room relaxed both men. Jack's was a short one as he was heading home after he dropped off Hank.

PUT THINGS BACK

Jack woke up in the morning with an idea for the grade six classes. He had to call John Henry in Ottawa and get his opinion.

Jack, "Hi John, this is Jack Wilson. I thought I could find you at work. How are you doing?

Jack ran his grade six rain drop and ribbon project by John. John was in agreement. Teenagers have the most at stake and they are also the most committed. They would have the energy and the networks to make it work.

John, "Jack I feel this is one of your best ideas."

In Ottawa, Saturday morning arrived and the very large yellow bus unloaded a passel of students with their teachers and parent chaperones. The university students walking by pretended to ignore the gathering, but Dougiee said hello to everyone that passed by. They couldn't ignore his eye contact and smile. To a student, they came out of their world and they all said "hi" or "hello" back.

The Tour

Each year John would take the elementary students on a campus tour, showing them large buildings, big lecture theaters, labs with horror film equipment working away, and the library, which was always a big hit. The university library was larger than their entire elementary school.

This year, John's tour went well and no one got lost. A couple of times one of the parent chaperones had to go back and get Dougiee as he was busy chatting to a student doing research. In another case he was feeding his packed snack to the animals in the science prep room. Dougiee would have hours of interesting stories to tell Grace and Mr. Bob when he got home.

The last stop before the lecture hall was snacks in the cafeteria and the all-important washroom break. For the young kids it was recess break in a very large cafeteria.

As Jack commented when he saw the student schedule, "Once you're fed and watered you're ready to listen."

Once the all-important head count was taken and all the chaperones had accounted for their charges, the group was off to the lecture theatre.

Unlike the sophisticated university students that liked to sit at the back, the front seats were at a premium and were filled first by the six graders. They did not have the experience to be cool.

John introduced Jack Wilson. The students were surprised that Jack grew up in Norwood, their town. They would have to share Norwood with this other person. Growing up in a small village and knowing everyone, young people often feel it's their own personal village. As age and experience settles in they realize that other people grew up in their town before them. They learn that many other older people also consider their town, their town. Norwood has a population of thirteen hundred and the Norwood fair on Thanksgiving weekend draws over fifty thousand people. A large percentage of those fifty thousand consider Norwood their town.

Jack was in his first day sweats, like the first time he had stood in front of those bright university students almost nine months ago.

Jack, "Good morning." He started to sketch the diving board at the Mill Pond in Norwood. The willow trees, the steel wheel, and the falls were quickly outlined. "On hot summer days how many of your parents swum in this pond? Let me say that again. How many of your parents swam in the pond?" Only a few students noticed the difference. They were busy remembering. Just about every hand went up.

Jack, "Do you know the pond that has the highest elevation, on the north side of highway seven, is called the Mill Pond? The pond on the south side of the highway is called the Upper Pond. Here's a question for you. Which pond drove the water turbines for the grist mill which was on the south side of the highway?"

Jack drew the old wooden grist mill and the farm wagons loaded with bags of grain to be ground. The horses were all standing still, tied up to the rail on the shady north side of the mill. "Before that mill was built the farmers had to take their grain to Cobourg to be ground. One man carrying one bag of grain could walk to Cobourg and carry the bag of flour back in one day. I am sure local farmers going by in wagons, carts, or sleighs would have given him a lift even if it was only a short distance to their turnoff."

Jack looked up at the students. "The local Norwood farmers were very happy to have that mill. It was later converted from water power to a one cylinder diesel engine. After that mill burned down the Blatchford feed mill was built along the south side of the railway tracks, just north of the cemetery. It also burned down, but years later. You know grain dust is just about as explosive as gun powder. You've got to keep things clean."

Jack quickly sketched two other favourite swimming holes along the Ouse River that were secret skinny-dipping spots and the students were now hooked. This man named Jack had passed the test. He was definitely from Norwood.

Two children were drawn lying down on their towels on the cement beside the water running over the falls. The old glacier spillway is still there today.

Jack looked up and asked; "How does your body cool down when you're not in the pond?"

Jack drew an arm and then the bubbles of sweat starting to appear.

The next drawing was a cross-section of the layers of skin and sweat glands in full board size.

Jack, "I need two volunteers to colour in these pictures." All hands went up and in a few minutes ten or twelve students were up at the board with coloured chalk, working and listening. Every three minutes Jack would ring the bell and another ten students would have their turn.

Jack remembered the advice he had been given by Hank: "Keep them busy and draw their knowledge out of them."

Ricarda wondered why all the different sized stools and two small step ladders were lined up against the side wall. John's brother, Spike, had collected them from around the campus and had stacked them near the wall in the lecture hall for the grade six visit.

Dougiee was the first up to the board and he soon fell in step with Jack. He picked up the brush when it fell, passed out the coloured chalk, rang the bell, and set the timer. Dougiee had learned to live with Mr. Bob and help him out as Mr Bob aged.

Thirty minutes slipped by, then sixty. Later, no one wanted to believe the clock when it turned eleven.

As Jack and the grade six students were working at the front on the board, the lecture hall quietly filled with university students. Many had the privilege of seeing Jack before and they were curious to see what was going on. How do you put a sign on a lecture door **"keep out"** when learning is going on? The university students sat and watched all the children and Jack, lost together in their own world. The young students were absorbed. In two hours the students had a good idea how the body cools itself, the history of shade, protecting your skin from sun rays (a lot of yellow chalk was used on that diagram), and the importance of knowing your skin type as well as how to protect yourself.

⁓

When Jack asked the grade six group to put up their hands if they lived on a working farm, a number of hands went high in the air. Duncan answered Jack's question and explained how their family farm rotated corn, soybeans, oats, and winter wheat in a four year cycle. Jack sketched the crops as Duncan was explaining to the class. Dougiee kept handing Jack all the different coloured chalk as he needed them.

In the second hour, Jack covered the crop rotation system in place in Norwood and also how the earth keeps cool. The world's oceans, lakes, and rivers sweated (evaporation). The grade six students soon realized that everything living on planet Earth had to have a system of keeping cool. Humans had skin that sweated: Dogs had feet that sweated and long tongues that dripped water to cool down the blood. Evaporation and Shade were the two big air conditioners. Trees were the second most important air conditioners on the planet. Large white pine trees and towering red oaks came to life on the board, and evaporation through the leaves and needles were blown up on separate diagrams. There was no question the students could feel the cool woods as Jack walked them through and around the trees in his blackboard forest. The next board showed the same trees dressed in snow with a white blanket on the ground. The sun could not penetrate down to the ground and as the snow fell off the branches, inches then feet, of snow building up could be seen. Spring arrived and the sun came out. The tree tops warmed, the roots started to come alive, and the snow around the tree trunks started to melt. Water seeped into the topsoil then soaked down into the roots. Who would be the lucky one to colour in those trees and undergrowth?

Jack sketched the huge maple tree on Queen Street. Every student recognized the large old maple. Jack pulled out a piece of orange chalk and drew a large circle with the enclosed slanted line on the big trunk. Jack could hear the students sucking in their breath as the young audience knew this was going to be another mistake.

Jack turned to his audience. "Someone has decided that it would be less expensive to cut the tree than cable the large limbs and let it live for another forty or fifty years. Maybe we can come up with a plan to save that old tree."

There was a cheer from the young students. Jack thought to himself, maybe the rain drop program will be the answer to save our old trees and get more trees planted.

Jack remembered Hank's advice: IPSA (Introduction, Presentation, Summary, and Application) "Don't give them homework, get them to do something and this will reinforce their learning."

The oceans, seas, lakes, and rivers are large evaporation trays. The forests of our world are the second largest air conditioners for the planet. The amount of water that evaporates from the forests has been reduced over time and the world is heating up. In the spring, when the snow melts and the water runs off the farm fields, the earth warms up. The forests are still full of snow and the ground is frozen. Forests keep us cool and we have cut them down.

Jack turned to his young audience. "When I was your age Saturday mornings was for play. My buddies and I would break off large chunks of ice along the Ouse River banks, by the bridge to school, and ride them until we fell off. Not sure if your parents would allow that today. The spring flooding of the Trent Severn Waterway is made much worse by all the empty farm fields not holding the water and many low lying fields have government sponsored drainage tile so they will hold even less water in the spring."

Wetlands

Jack had the young helpers clean off the boards. He printed "Wetlands" at the top of the centre board. Starting on the far left board he drew a wetland full of small trees, bulrushes, a few birds, and a weasel poking its head out with a mouse in its jaws. Then he showed the machinery scraping off the vegetation and digging trenches. Rolls of plastic pipe to drain the soil were stacked around the field. Jack spoke softly into his microphone, "The Ontario government that helps finance this university building, and your schools in Norwood, also provided the grants for this farmer to drain this wetland. Not only are we losing the natural vegetation and wildlife we are also losing an important catch basin to store excess moisture when the snow melts in the spring. Wetlands suck up excess moisture quickly. They act as natural reservoirs to prevent flooding. Wetlands allow water to seep slowly into the groundwater table, and increase evaporation to cool planet Earth. We need wetlands."

Jack quickly sketched two more wetlands. On the first wetland he placed a constuction yard. "This yard is on the north side of Highway 7, west of the Keene turnoff. A company has a corn-fed ethanol plant near Havelock. They are now buying up farms, stripping out the fence rows, and growing genetically modified corn to produce ethanol. It's hard to turn the big equipment around in small fields. The federal government chips in eight million dollars. It will soon be thirteen million dollars and the provincial government gave the company $772,000 to expand. More trees more wind blocks, more depletion of our farm land financed by our two levels of government. Providing work for the locals and building up the bank account are priorities over our planet."

Angie Akins was squirming in her seat as her mom and dad both worked for the company. Her dad operates heavy machinery and for the last couple of years, has been the foreman cleaning out the fence lines on the farms the company was buying up to produce more and more corn. Her dad, the first generation off the farm, mentioned at the supper table that growing corn year after year would eventually ruin the soil. She remembers her dad saying that farmers just can't put that many chemicals on the land year after year.

⟨∿⟩

The second wetland, south of Lansdowne Street off the parkway in Peterborough has a coffee shop and a tourist building. Jack, speaking softly again, continued, "You know we used to call these areas swamps and no one wanted them as they had no economic value. If you look at the historic maps at the Peterborough Museum you'll notice most of southern Peterborough was once wetlands. Looking at the young audience, "You must know of other wetlands that have been filled in." He asked them to use the centre board to write down the locations. Dougiee was the first. He printed, The 9th Line around Rotten Lake. About half the wetlands posted on the centre blackboard were concession roads and county roads. Allen wrote Orono, the LCBO, grocery store, and the soccer fields.

To Jack's surprise Dougiee picked up the chalk when everyone had finished and wrote, 'Corduroy roads were built to go across wetlands and swamps". Jack knew most adults would not know what a corduroy road was or how it was built. Jack circled corduroy and then quickly sketched how the road was built over and through a wetland.

Jack ended the wetland story with Trent Severn Waterway. "All of you in the audience live in the watershed drained by Trent Severn Waterway. The Ouse River drains into Rice Lake. Guess what? Their Peterborough head office and storage yard is a filled in wetland on the shore of Little Lake. The complete area on Armour Road is a wet land that has been filled in. Guess who gets flooded out next?"

Jack, as he was talking, outlined the city map of Peterborough and coloured in the southern half of the city with blue chalk.

Jack turned to his audience. "At Christmas time have you noticed how much water your tree will drink? It's a small tree in comparison to a full grown mature tree. Imagine how much water a big tree can hold and drink in a day. All that moisture sucked up by the roots eventually evaporates through the leaves, needles, or tender branches. Water vapour condenses, forms clouds, and cleans the air when it rains. Precipitation cools the planet. Trees and vegetation purify our water. The farm students in your class will have noticed trees that grow near a manure pile. These trees not only grow faster, but all the groundwater they absorb is cleaned and is returned to the earth's surface downwind, clean as a whistle."

As Jack was talking he drew the water moisture leaving the leaves, forming clouds and raining downwind.

When he drew the world map and showed them the vast forests around the world that had been cut down, the students were very quiet and sad.

Jack spoke into his microphone. "French oak forests were cut down to grow grapes; mountain sides were grooved to carry the water away quickly to bring on the heat and conditions necessary for wine. Germany, Italy, Spain, and Portugal have cut down their forests to grow grapes. Have you been to Niagara Falls? The Niagara peninsula was a forest years ago. Every country in the world has cut their trees down, not knowing they were destroying the world's air and water conditioners. Forest removal has raised the level of the oceans. Forests around the world have been cut down to grow crops, pasture animals, and to pave roads. We also cover the land with buildings.

If you have a chance to go to south-western Ontario you'll see farms with one hundred acre fields and not a tree in sight, not even trees along the roadway. You'll occasionally see a tree-lined driveway but that's it. On the back of the farm you'll see a small maple woodlot that traditionally provided the farmer with their first cash crop of the season and firewood to heat with. Sugaring off traditionally paid for the seed to plant the crops. When it snows in Bruce County the roads are closed due to blowing snow. Bruce, Grey, Huron, and Wellington counties are probably the best examples of what not to do. Should we tell them to plant some trees?

Governments around the world own most of the highways. They have clear-cut strips around the globe. All levels of government together are the largest clear-cutters in the world; together they have clear-cut the most trees on our planet Earth. In Ontario our road allowances are one chain wide."

Dougiee printed on the board, "I chain = 66 feet"

Jack drew the cross-section of a highway and showed the buildup of fill, layers of stone, then gravel, and finally the top coat of water-proof black asphalt. The ditches on each side of the two lane road were sketched in proportion. The students could clearly see the empty banks before the farm fences. Jack then drew in three rows of trees on the government side of the fence and three rows of trees on the private side of the fence.

To test Dougiee, Jack asked the grade sixers, "How many trees could you plant in a mile if the trees were five feet apart?"

Dougiee did the arithmetic on the board.

$(5280 / 5 = 1056)$ x 6 rows = 6336 x 2 sides = 12,672 new trees.

Jack then added, "How many miles of trees would you have to plant to plant one million?"

Dougiee put down the chalk briefly and pulled out his calculator.

Dougiee put on the board 158 miles $/2 = 79$miles

Jack was impressed again, as even he had forgotten for the moment that a road has two sides. Jack said to himself, "I have to visit Grace and Mr. Bob."

Dougiee, before he rang the bell, wrote on the board. "Tree seed shortage."

John was getting to like Dougiee more and more. He said to himself, "How does a twelve year-old boy know what a chain is, let alone how long it is?"

Norwood Esker

Jack drew the esker in Norwood, put on the crinkled water tower, and cenotaph.

The crinkled water tower was a result of a group of pranksters opening up a number of fire hydrants all at once. The water flowing out of the hydrants was greater than the air going in the vent at the top of the water tower. As the inside air pressure decreased, the outside air pressure pushed in the thin metal walls like a crushed pop can. It was discovered later that bird nests had almost closed the one and only vent at the top of the water tower.

Behind the war monument was the pump house that filled the water tower on top of the hill. Gravity created the water pressure to supply all the town water. On the cenotaph diagram, Jack wrote the first four names and asked any student in the class to write down the names of any relative that was also on the monument. He drew in the original small ponds around the esker and slowly wiped them off the board as the town continued to pump water out of the esker. The large swamp, now called wetland, that stopped concession Eight from continuing north was now a very small area.

Splash not Timberrr!

Jack ended the class with a large white pine tree being felled by a young logger. Jack coloured half the tree blue and then he wrote, "It's not Timberrrr it's Splash."

Jack leaned into his microphone and paused for a couple of seconds. It's one of those moments when you feel the speaker is letting you in on a secret. "Trees are over fifty percent water and they hold the largest amount of fresh water in the world."

Jack passed around two pine boards, one green freshly cut, and the other kiln-dried down to six percent moisture content. The two other samples were red oak. The boards were one inch thick by twelve inches wide by twelve inches long. "You can feel the difference in weight. The heavy board has just been sawn from a green log and the lighter one has been dried for six months. The moisture is gone from the lighter board. Humans have cut down eighty percent of all the trees on our planet Earth. Gravity has pulled all that moisture back to the oceans. Notice how much smaller the dry board is compared to the green board. As the water leaves the lumber, it shrinks."

Dougiee, "Mr. Jack, those boards are one board foot and one-twelfth of a cubic foot. A cubic foot of water weighs sixty-two and a half pounds, or over twenty-eight thousand grams, or twenty-eight kilograms. Twelve board feet is a cubic foot and if trees are fifty percent water then a cubic foot of tree has thirty-one pounds of water or fourteen kilograms of water."

Jack, "Dougiee, put that on the board and circle it please."

Dougiee slipped his calculator into his pocket and approached the black board with a piece of blue chalk.

No one was surprised to hear Dougiee quote the numbers. They just knew he had a different teacher and a different skill set. Being around Mr. Bob day after day, you learn a lot.

Dougiee then summed it up, "Mr. Bob always told me to put things back. They have to put the trees back and cool the planet down. We can lower the oceans if we all work together."

Jack: "Dougiee, you're my assistant; I could not say it any better."

Dougiee; "Professor Jack, I have one question."

"Yes Dougiee, only one?"

Dougiee: "If trees are fifty percent water, why do we have forest fires?"

Jack: "A very good question. Let me try to explain."

Allen had his hand up. Jack nodded. Allen, "Trees are fifty percent carbon and the carbon burning generates enough heat that it will boil off the water and the tree carbon will burn. A single tree will not burn but in a forest fire the combined heat will do the job."

Julie and Barbara looked at each other with a questioning look.

Jack turned to the board and drew two kernels of popping corn. "Popping corn has to have moisture to pop, and then you add high heat. On the board students could see the stove top was red under the corn. The steam was building inside the kernel, then it exploded on the board.

I'm sure many of you have forgotten to prick a potato, placed it into the oven, and came back later to see that it had exploded.

If you talk to forest firefighters they will tell you one of the most deadly hazards is exploding trees. Trees will sometimes get super-heated if the wind is pushing the heat ahead of the fire. Depending on the type of tree and the amount of heat, the tree can explode. The tremendous heat of lighting can have a similar impact on a green tree. Thousands of slivers explode outward from the tree. If you are close to a tree when it pops, well, you'll either look like a porcupine or your friends will be getting you later. Everything burns when it gets hot enough; even aluminum subway cars underground in a tunnel will burn. Forest fires are so hot they boil off the water and the burning carbon produces more heat to continue spreading the fire. Wind and convection currents will keep moving the fire until it hits an area where there is no fuel, or a rain storm puts it out."

Jack continued, "On a cold day in the winter you will notice that the smoke from a wood stove or wood-burning furnace will go almost straight up. What you are seeing is mostly steam. The water left in the firewood is boiling off. Creosote is created on the inside linings of chimneys when the water content of the wood is too high. The heat from the burning carbon is not enough to burn all the wood or boil all the water. As you know, this creosote can build up when you're burning the fire too slowly or you're trying to burn green wood."

Jack once again remembered Hank's IPSA. He thought to himself, I've done the introduction and presentation. Just the summary and application left."

Jack; "Let's review"

Jack printed on the centre blackboard.

COOLING OUR PLANET EARTH

SEA

"S" is for shade

"E" is for evaporation

"A" is for altitude

SHADE

Jack spoke slowly into the microphone, "There are only three ways for our planet to cool down. The first is shade."

Jack sketched four circles on the board. The yellow sun was placed in the centre of the four circles.

"The tilt of the earth in the rotation around the sun causes the seasons and the polar ice caps." Jack drew a line through each of the four circles representing Earth's axis tilted on the correct angle (23.4 degrees)

Jack posed the question the grade six students, "How does the tilt create the seasons?"

Allan had his hand up and Jack nodded.

Allan: "When you sit close to the campfire and roast marshmallows on a cold night, you roast you face and freeze your buns. Then when you turn around with the marshmallow and warm your buns your face gets cold. That's summer and winter and half way through the turns you have spring and fall.

Allan, The turning of the earth allows the sun to heat during the day and the shade to cool the planet at night. At the North Pole during the winter there is just about no sun and only shade and it is very cold."

Jack drew the ice cap on the North Pole and then on the South Pole.

Jack, "How fast is the earth turning?"

Dougiee had that answer and the class waited. They knew he would have it in his memory bank.

Dougiee, "The Earth turns around once, that is, it turns on its axis once in twenty-four hours and the circumference of the earth is 24,901.55 miles (40,075.16 kilometers). At the equator we are moving at approximately one thousand miles an hour, from west to east. If you stood at the North or South Pole you would only turn around once in a day. This is why our weather usually comes from the west. The earth slips under the atmosphere. Mr. Bob says our atmosphere is like the grease around a wheel bearing. It allows the planet to keep revolving because if we slow down we're in trouble."

Jack nodded and smiled. He caught John sitting at the side of the lecture hall with a smile on his face.

EVAPORATION

Jack paused and looked over again at John and the two teachers from Norwood, Barbara and Julie. They were making room for Ricarda to join them.

Jack nodded to Ricarda and continued, "Evaporation is the second cooling mechanism. Evaporation of water, oceans, seas, rivers, lakes, the Ouse River-all vegetation, especially the big plants called trees.

How much water is there in a tree?" This was to be a rhetorical question but Allan had his hand up.

Allan: "Most trees are half full; I mean they are fifty to eighty percent water. Oh yeah, I mean the weight of a tree is fifty to eighty percent water.

My dad and Uncle Malcolm are members of Ducks Unlimited. Ducks Unlimited know if you don't have wetlands you don't have ducks and there would be no duck hunting. I'am going to tell them about how wetlands allow our planet to keep cool."

Barbara turned to Julie and whispered, "I just can't believe Allen answering questions. He's so quiet in class and he never puts his hand up."

Julie: "Now you sure know what he's interested in."

Allen continued, "Eruptions, when the earth opens up and belches out dirty clouds, we get shade and if it's big enough, we get an ice age. We are not allowing planet Earth to cool itself."

ALTITUDE

Jack, "The third and final way our planet can cool itself is all related to altitude, not attitude."

The university students at the back of the class stifled a laugh as it went over the heads of the younger crowd up front.

"If you have travelled with your parents or had any other opportunity to go high up in the mountains, you will have noticed that it gets cooler as you climb higher. Raise your hand if you have ever seen snow capped mountains in the summer."

Two or three hands went up.

Jack, "Tops of mountains are cold. When water freezes it expands. That's why ice floats on the more dense water. Ice is lighter than water so the heavier water underneath pushes up the ice. When moisture on the surface of the planet evaporates, over a lake, ocean, or land mass, the moisture in the air is lighter than the surrounding air and the heavier surrounding air pushes up the moist air. Gravity is the force causing the lighter ice to float and the lighter moist air to rise into the sky. Hot air does not rise by itself. There is no force pulling up hot air. Heavier air is pushing up hot air. The weather folks call it air pressure. I think of weight and gravity, not air pressure.

The evaporation of water into the air creates the clouds. Many clouds rise as high as mountain tops. The water moisture cools as it is being pushed higher and higher."

Jack works busily on the board drawing the mountain, clouds, and water molecules forming around dust particles, then drops of rain starting to fall. As Jack moves from board to board, the colouring team fills in behind. Jack does not even hear the sound of the bell anymore as noise just blends in to the action on stage.

The students change and a fresh crew starts to busily colour in the sketches.

Jack continued, "When the water moisture joins with dust particles they eventually fall, and this is called rain. As gravity pulls it down out of the sky the rain passes through the warmer air, cooling down the air. In turn, the warm air warms up the cool rain. Ice balls or hail are created when a heavy current of air pushes these water droplets higher thereby making them much cooler. They can freeze into small round balls. If the uplift of air continues pushing the small ice balls, they get larger and larger, until finally the weight of the ice is too much and gravity pulls them down to the ground. Also in Canada, we know that water moisture in the air cools and forms frozen crystals."

Jack quickly drew three very large snow crystals.

"Snow cools down planet Earth. Snow accumulation on mountain tops can be compared to the iceboxes of years ago. During the winter months the snow accumulates and the white hat on the mountain tops grows larger. During the warmer months the snow and ice melt and the cool mountain streams and drafts of cool air sliding down the mountainsides cool our planet.

A hand went up from the audience of grade six students. Allan had his hand up and Jack acknowledged.

"Mr. Jack, in the mornings in the spring and fall, why are the grass and driveways wet? In the morning later in the fall we have frost on the truck windows?"

Jack did not have to answer as Dougiee turned and said, "That's an example of shade cooling the planet. When the sun goes down, because earth is turning, the moist air near the surface cools and gravity pulls the water moisture back down. In late fall as the number of hours of the heating sun get shorter, the water freezes. Frost is just cold dew."

Jack, "I couldn't have explained it better."

Julie turned to Barbara, "What are you going to do next year when Dougiee moves onto the next grade?"

Jack drew a teeter-totter. "No force is pulling us up on planet earth, only gravity is pulling us down. You can weigh as much as you want and you will not lift up. As soon as someone heavier is on the other end of the board, you are going up. Air all around us works the same way. Heavier air slides along the surface of our planet pushing up the lighter air." The two characters on the teeter totter looked very much like the smallest and largest students in the audience.

Allan's assigned travelling buddy, Sean Sullivan, had his hand up.

Jack nodded. Sean, "Last summer our family took our first trip ever back to Ireland. In the plane there was a screen on the back of the seat in front of us and I watched it a lot. We were flying above the clouds and in the clouds. The temperature outside was minus fifteen and minus twenty, sometimes colder. Our trip was in July and was it ever hot when we landed."

Jack recognized the information with a smile.

Barbara glanced at Julie. "Looks like we know what Sean will be talking about for awhile."

EVAPORATION

Jack, "It is very important that we have evaporation. Lakes, rivers, wetlands, vegetation-especially trees-are very important in cooling our planet. The heat of the sun is absorbed when it evaporates the water. Evaporation cools the planet. When the water moisture in the air is pushed up, like to the top of a mountain, it cools. When the rain falls it cleans the air as well as cooling it. I sometimes repeat myself. It is so important that we do not stop the planet from sweating. Rain cools the planet. Cutting our trees, filling in our wetlands, filling in our lakes or oceans are all contributing to heating up our planet. Farmers get paid by the provincial government to drain wet fields that should have been left in forests to begin with. Cities like Peterborough have allowed wetlands to be filled in. A new drive-through fast food outlet and even the city tourist welcome centre are now sitting on fill where there once was a viable wetland.

Ironically these buildings are all air conditioned burning up more electricity to keep the staff and customers cool. As you know, operating air conditioners adds to the heat of our planet. Actions like this are a double whammy! You destroy the natural way the planet cools and then construct buildings and air-condition them."

Jack had that twinkle in his eye. "What is cooler in the summer, a large black asphalt parking lot or a wetland filled with bulrushes, small trees, and marsh plants?

Jack looked at Allan. We should all be members, or at least support Ducks Unlimited. I'm not a hunter but if it was not for that group we would have fewer wetlands and fewer ducks than we have today."

EARTH

Jack started sketching a hot desert.

"Think about the hot deserts of the world. There are no lakes, rivers, or trees. They have little to no vegetation most of the year. There is no water for evaporation so there are no clouds forming, and thus no rain. Only when the winds blow clouds over the deserts and the clouds rise and cool do the hot deserts get rain. Do you know that when it does rain in the desert it catches people by surprise? Many people have drowned in these flash floods, as the deserts do not have ways to store and soak up the water quickly. The hot deserts of the world are cold at night, as there is no water moisture in the air to act as an insulating blanket. Bake during the day and freeze at night-much like being on the moon, but not so extreme because the hot deserts are not a big part of the Earth's surface. The cold deserts of the world like the Arctic and Antarctic have little to no evaporation and few clouds and therefore little chance of snow let alone rain.

Jack continued, "Here's a question for you. If you look at the hot deserts of the world and you start to irrigate the deserts from the windward side and plant vegetation, what would happen? Think about growing food and having the vegetation releasing moisture and providing the building blocks for clouds. The money paid to Middle East oil families could be used to provide food and water and thus help cool the planet. Desalination plants would provide the water. They know how to pipe oil. What about piping water?"

Governments are a big part of the problem

Jack, "The Ontario government, like many governments around the world, even today, pays farmers to put in drainage pipes to drain wet fields. They are paying farmers to increase the ocean levels of the world. If you ever take the time to study the drainage systems they've put in place in the U.K., you'll be shocked at how much water they're dumping into the oceans.

꩜

Remember home and office air conditioners heat up the earth. You have to use energy. Put your hand at the back of a refrigerator and feel the hot radiator. Inside the box it's cool and cold in the freezer section, but the heat generated is more than the cool temperatures inside the insulated box. Every refrigerator, every freezer, every walk-in cooler, every air conditioner puts out more heat than what it cools. In a large, densely packed city, when the heat arrives in the summer and the air conditioners are turned on, the outside temperature heats up and the air conditioners have to burn even more energy."

Jack used an example to demonstrate the city heating up when air conditioners are turned on. "Put a refrigerator in a room and turn it on. Electricity evaporates the freon, the motor adds heat to the gym, and the contents inside the box cool. If you then filled the room with refrigerators, you would have one hot room. At least you could have a cool drink when you pay the energy bill.

Car, truck, house, office, air conditioning,... then he dropped the bomb on the young hockey players in the crowd, artificial ice in the summer. We're heating up the world for pleasure and fun."

The young students suddenly realized that everyone in the future would have to make hard decisions.

Air conditioning killed the Ontario Tourist Industry.

Jack, "Fifty years ago the Kawartha Lakes around Norwood attracted Americans by the thousands. You would see Ohio, New York, or Pennsylvania licence plates on every other car. This was a time before car or home air conditioning. The one way you could get cool during the summer was to go to the northern lakes and rivers. People built cottages and summer resorts and you booked a year in advance to rent these cool spots. Once air-conditioning became affordable, there was no longer a need to go north to the lakes and rivers. Americans stayed home. The second hit for the tourist industry was cheap air travel. Why take your holidays in the summertime when you could fly south and have a weather guarantee? It's cheaper to fly south for a week, and eat and drink, than it is to stay here in Canada during the summer with no guarantee of the weather. Air-conditioning is helping to heat up our planet."

Homework

How do you explain trippage to grade six students? Jack came up with the idea that he would ask the students to write him a letter.

Julia, "Jack don't get your hopes up. Students today text-message and facebook. They don't like to put a pen or pencil in their hands, let alone write.

I am not sure they know where the post office is, much less ever having been there."

Jack, "Sometimes we like to step back in time and do something our parents or grandparents did. My uncle Rob started a contest with the family and he challenged us all to grow the largest tomato. The adults all put in ten dollars and the kids all put in one dollar. There was over one hundred dollars in the pot and they had weigh-in on August 13th, Gary's birthday, he was the youngest!

Julie, if you don't mind let's see if your class will write me a letter and I promise I will write them back."

Homework - this is the killer for many students. Jack had talked to his son, Chip, and also received sound advice from Hank. He had a plan.

At the end of the session Jack turned to the audience and said, "You have no homework." All the students looked up and cheered.

"You are not off the hook though. I want you to help me!"

Hope Friday was sitting at the back of the lecture hall and he turned to his girlfriend and whispered, "Jack is reeling in more help for his projects. He's a master at this. Jack is a fox."

It seemed like a long pause but it was purposely timed.

Jack, "I need your help and only you can do it. You have to spread the word. This is an idea but you are free to change it. If we could mark the trees in Norwood, especially the large trees along the sidewalks and streets with a system to let people know how much water is in each tree, it would be a start. We have to get people's attention.

I think you'll find the largest trees on Queen or King Street. Every house that has a working cistern should be identified and recognized.

Call or write to your cousins and ask them to spread the word and start the Norwood tree program."

Dougiee: "You know, Mr. Bob told me how Norwood got its name. Norwood is short for North Woods. Mr. Bob laughed and said he was like the locals bending the truth. Norwood was named by Harriet Keeler in the post office. She was English and named it after a suburb in England. Mr. Bob will likely say, Norwood should start the movement back to reforesting the world. Mr. Bob thinks big.

I am going to call cousins Bruce and Margaret in Oakville. They could start putting ribbons on the large oak trees. There are only a few left.

Mr. Bob also says the pioneers cut the oaks down and more importantly forgot to replant. He also says that Oakville should be called Fordville because cars and profits are more important than oak trees. Mr. Bob has a Dodge Ram bought from Stewarts in town."

◦◦◦

Jack nodded to John and John came to the front of the lecture hall with Julie, Barbara, and Ricarda. John thanked Jack for his lecture and diagrams and the young audience applauded.

Julie instructed the students to be sure to pick up all their clothing, belongings and any paper or garbage that dropped on the floor. Her last words were; "Stick with your travelling buddy and we're meeting outside after you have a short washroom break. Girls, can you beat the guys out of the washroom this time?"

Barbara announced to the group that Jack had prepared a clipboard for every student with their name on it. "Please pick up your clipboard as you leave the lecture hall." Barbara did not have to worry about any student losing their new found treasure. Jack had also made sure they were in very bright colours.

Jack opened up two big cardboard boxes near the door and asked Dougiee if he and Allan with some friends could pass the clipboards out. The name of every student was in big letters on the top of their own personal clipboard and they were in alphabetical order. The title on the back of each board was:

CHANGE TIMBERRRR TO "SPLASH" AND CHANGE THE WORLD

On the back of each clipboard was a chart. The chart had three columns: The tree type, the board foot weight, and the pounds/kilograms of water per board-foot. At the bottom of the chart was a simple method of calculating the number of board-feet in a tree.

⁓

Jack and the grade six students went outside and he showed them how to measure a tree with a pencil and how to calculate the volume of water.

He picked Susie, the shyest student, to hold up the pencil at arm's length. Jack demonstrated first, then turned it over to Susie.

"Susie, hold on to the pencil so the rubber is visually touching the ground and the tip of the pencil is at the top of the tree. Close one eye. Now rotate the pencil sideways, keeping the rubber on the ground in front of the tree." He had Greg walk over to where the tip of the pencil visually touched the ground. Greg then counted his steps to the tree. The distance was measured and they worked out Greg's stepping distance. Greg didn't need a tape after that; he knew how to measure the height of any tree with a pencil and his own steps.

"Greg, could you hug that tree?" In two and a-half hugs, Greg was around the tree.

Greg's arms, hands, and full hug were measured with the tape and Greg could now calculate the circumference of trees. All the students took turns measuring their travelling partner's step distance and arm length. Jack referred the arm length as their wing span. The chaperons took their turns with the tapes. They also wanted to measure their steps and arm spans. Jack thought to himself. "This plan just might work."

Jack: "Don't forget, your wing span is the same as your height. Since you guys are growing every year you'll have to increase those numbers."

Julie and Barbara gave the signal and the pack of young ones swarmed towards the bus.

Beside the school bus, Jack talked to Julie and Barbara as they checked off the students and chaperons from their class lists on their clipboards. Watching the young children as they climbed the steps, he overheard two boys talking to each other. Jack received the ultimate compliment. "University is fun. When I get older I want to go."

The sixth grade tour was heading to the parliament buildings to meet with the local Norwood area Member of Parliament. He had arranged to provide a picnic lunch for the complete group. The MP was so embarrassed by how the elected representatives behaved in parliament that he would try to schedule class visits around session times. He was proud to be their representative. Members of parliament often behave exactly opposite to how students are taught to behave in class. Shouting, insulting, and talking over one another were not good examples to set, so sitting sessions were avoided.

He referred to his elected colleagues as Outlaws in Ottawa.

Back Home

At first the villagers in Norwood were more than a bit curious when they saw young kids hugging trees and putting on blue ribbons. The young students were quick to point out what they were doing and how important it was to let everyone know that trees were full of water.

A very large maple on Queen Street was marked by the town to come down. After the ribbon was tied on it, the town work crew decided that three cables strategically placed could keep the old senior together for a few more years. The orange circle with the cross took a few years to fade but it reminded everyone that is was spared. Roy, the town manager, knew exactly where to drill in the anchors and place the three cables. Two young sugar maples were planted on each side to take up the future challenge.

Norwood was the first village to ribbon their trees and start the movement of reforesting the planet-no small feat for a small town with two grade six classes willing to take up the challenge. The ribbon campaign spread rapidly.

Dougiee, Allan, Sean, and Greg would go on to be life-time friends. Their relationship was cemented in Ottawa. On the trip home they were allowed to sit together in the back of the bus. At the end of the three hour ride back to Norwood they had a plan to help Professor Jack.

Part of the conversation on the way home in the bus went like this: Dougiee: "Allan, you're a 4-H member. Do you think the other members of your club will get involved in planting seedlings?"

Allan: "Farmers, all they talk about is the weather-not enough rain, too much rain, not enough sun, too much sun, and the list goes on. My friends would be very interested and they have the land to grow the seedlings. 4-H stands for head, heart, hands, and health. Most farmers think trees are a nuisance. They shade their crops and suck up ground moisture and the big tractor air-conditioned cabs get scrubbed and scratched when they hit tree branches. Farmers are the worst. They cut down everything. They have to produce more for less every year and they protect their crops at all costs. Most farms have a small woodlot and they don't even join the local woodlot associations. It'll be hard to get them interested. I have an idea. We all had to raise a steer one year, and the next year we had to have a sow and raise a litter of piglets. Maybe I can suggest we all grow a hundred, no a thousand, seedlings." (And this is how it all began)

⁓

Barbara and Julie spoke on the bus on the way home.

Barbara, "You know you can always tell when you meet someone who doesn't work with children in a school. Jack thinks the grade six classes can start this project and get it rolling. It'll take the commitment of the whole school."

Julie, "Funny you should mention that, because in our school, nothing happens unless Veronica, our new principal approves it. This is her second year as principal and since the day she was hired she has had her hand on the next rung up the ladder. Every activity in the school had better line up to her personal goals or it's not going to happen. I think I can pitch this project in such a way that she can be seen as a leader to her superintendent and trustees. If we can get her on side, it will happen. She's the best principal we've had but she won't be with us long. She's on her way up and she's in a hurry.

To be allowed to take this trip to Ottawa with your public school, was a sales pitch. The new board initiative of col-laboration with our partners came in handy. Veronica needed something to fill in that slot on her resume and this was her opportunity to complete the check-list. God forbid that Catholic and Protestant children should ride on the same school bus. You know in Toronto, the Transit Commission has a rule that Catholics sit on the left and Protestants sit on the right side of buses, subways, and street cars."

They both laughed.

Barbara: "Our principal is a "go by the books" kind of guy. He's not looking for a promotion; he's looking after not doing anything wrong. To get this project in place we'll have to cut down a forest to produce all the permission forms, with newsletters, and memos that are mandated by our board. When Joe first came to our school he was really concerned about the horse farm next door. Sue took him to one side and said, "Joe, eighty percent of our students are from the farm. They're not even going to pay attention to the stud horses next door. On the farm, Bobby the bull or the long arm of the vet do the same thing.

My saving grace is fast Eddie. He's our grade eight teacher and he has never heard of a rule book, let alone read one. He gets away with murder as he's the only one that can handle the tough students in the school. We had two bullies and they got to grade eight last year. Eddie tamed them in their first month. They never knew what hit them. Eddie had a plan in place before they walked through his door.

The twins now catch a high school bus after school-the bus driver is a friend of Eddie's-and they're dropped off on the eighth line at a dairy farm. Ronald and Mary are my uncle and aunt. They've never had children and they're close to retiring and selling off the milk contract, herd, and machinery. They'll go into cash crops for the balance of their years on the farm. Eddie gave them the opportunity of hiring two high energy girls to do the second milking."

Barbara continued. "Katrin and Carey, the twins, usually get off the bus, change into their work clothes, and head to the barn with a snack in their hands. Mary is a great cook. In two hours Ronald and the girls have thirty-five cows milked and bedded down for the night. The twins then come back to the house, shower and dress, and sit down with Ronald and Mary to a home-cooked farm meal. After dinner the twins have their own desks to do their homework on, supervised-and a private tutor named Ronald. Meanwhile Mary packs them their lunch for next day as she gets a treat ready for Annie. At eight o'clock Mary drives the twins home and then slips over to visit Annie and put her to bed in the nursing home.

◦◡◦

The town looks after Annie. Mary is all too happy to tuck her into bed each night with a big farm hug. In her younger days Annie was a domestic, an orphan sent over from England to a farm family. I think they're called Barnardo children, named after the orphanage that placed so many unwanted waifs. If you think we've treated our native children badly you have to hear the stories of these children, especially some of the pretty girls. There was no follow-up for the deserted orphans. They were just placed and forgotten.

Annie, at ninety-two would go to sleep and her long-term memory played back the hunger pains she experienced day and night as a very small child. She would wake up in the middle of the night feeling hungry but did not want to eat. Getting back to sleep was a problem and she would be found fretting by the staff on night rounds. Mary was in town one day catching up on the local goings on as she ran the errands to keep the kitchen and farm running. Her neighbour, Girty, was picking up her nursing home uniform at the dry cleaners and bumped into Mary. In their casual conversation Girty mentioned Annie waking up in the middle of the night hungry. They had tried everything, even double snacks at night time before she went to sleep. Girty also mentioned that Annie was gaining a bit too much weight."

Barbara continued the story, "Mary had read about the London children that were displaced during the Second World War and their fear of starving. She had an idea. Mary made a snack every day and wrapped it in wax paper, and taped the ends shut. After she gave Annie her big hug every night, she made sure Annie saw her snack and asked Annie to hang onto it. Annie started sleeping through the night with the snack in her hand.

"Annie was always attractive, so you can only imagine what happened to her growing up as basically an indentured servant. Annie passed away peacefully one night not so long ago. She had the wax paper-wrapped treat firmly held in both hands."

Barbara, "Saturday mornings the twins would meet Mary at the grocery store and after shopping, went off to the farm for the day. Can you believe this-on Sunday after Sunday school they would join Ronald and Mary in church and go home with them?"

Julie, "How did they clean up their foul mouths?"

Barbara, "I asked my uncle and he told me how he did it. You know I've never heard my uncle swear once."

Barbara related the story told to her by Uncle Ronald. "One evening in the barn a milker fell off a cow. The cow, Sara, had recently started a habit of lifting her back foot and kicking it off. Katrin took a brush and smacked it across Sara's back and the curse words flew."

Ronald caught his breath and lowered his eyes. "He walked over and thought this would be the time to give a bit of fatherly advice. He purposely put some manure on the tip of his boot. He walked over to Katrin and after she had the milker back on she looked at him, then his boots. He said, "Katrin," pointing at his boots. You can say that I have shit on my boots or I can say I have manure on my boots. One expression puts me in the gutter, one takes me out of it. Katrin, you have a decision to make.

As Uncle Ronald was about to leave and he could see Katrin's shoulders slump and he said, with that twinkle in his eye, "Hitting Sara with the brush got her attention, she doesn't give a shit what you say."

They both laughed, and from that day after, he never heard a swear word from either girl.

Barbara continued her story, "The girls are putting in fifteen to twenty hours of work a week. Ronald had them down to the Royal Bank and they set up two accounts each. One is for college and one is to spend. Ronald makes them spend part of their money. He says, "You work hard, you should enjoy the fruits of your labour." Mary loves to sew and you should see some of the clothes that they're wearing. That grade eight class turned from the worst group to the best in the school.

There is more to tell but I first have to fill you in about Eddie. I thought of all the teachers on staff, in my mind he would be the one to help."

Follow up after the Ottawa Trip

Barbara, "Eddie, I am sure, will help me with this project. I can see it now. He will have the lawns removed and it will become a grow-op for seedlings. The playing field will have a two meter border of seedlings. In the winter every window will be filled with empty egg cartons growing seedlings. Eddie will have every student involved."

Back at school, in the staff room, Monday morning......

Barbara, "Hi Eddie."

Eddie: "How was your Ottawa trip?"

Barbara, "It was great! I have to talk to you about a project. I'm going to need your help."

Eddie, "I'm just putting the final touches on the "Big Tomato" growing competition. I'm scheduled to see Joe after bus duty today. You know, dot the i's and cross the t's and collect the paper.

Barbara, "We need to sit down at lunch. I've got this seat work to run off before class and I'm also on bus duty in ten minutes."

Eddie, "I have a cycle route scheduled for the senior boys and girls at lunch today so I will see you outside in a few minutes at the bus loop."

At the bus loop, Barbara gave Eddie a quick rundown of the lecture Jack Wilson presented to the grade six classes and the assignment of planting trees. They both agreed that Eddie needed more information. He did not want to reschedule his pre-arranged meeting with the principal. Barbara and Eddie tentatively set up a meeting after supper at her place. Barbara would contact Julie. Eddie cycled just about every evening and he would drop in to Barbara's home. He would drive his old half ton tonight as he figured the meeting would run late.

⌒⌒

Julie and Barbara had taken detailed notes in Ottawa. After an hour of explaining all they could remember to Eddie, it was time for Eddie to ask some questions.

Eddie: "How do you know trees contain fifty or sixty percent water?"

Julie, "Eddie, you have a huge Christmas tree plantation. If anyone should know, you should. When you cut trees are they full of water?"

Eddie: "In the summer when we trim them, they are full of gum or sap and in the winter, yes, the stump is wet and frozen. It's for sure that the tree when it is discarded, it is as light as a feather. No matter how much water they suck up in the house over four or five weeks, when you stop watering them, they are dry and light. You all know how many needles fall off when you try to get a Christmas trees out of the living room and out the door.

Eddie continued, I can't believe that no one before, not even a scientist or environmentalist, has ever proposed the idea that trees are full of water. You would think that all the maple syrup producers would have caught on. There is no question that we've mowed trees down around the world. It'll be interesting to watch and listen to the experts, once they find out that the ocean levels are in direct relationship to the world's forests, and that global warming can be reversed by planting air conditioners called trees. Boy, I'm excited about this."

Julie, "There's much more to the story about ocean levels but let's focus on the trees first. That's something we can get started on. Remember, we're talking about the weight of the trees and the weight of water in the trees."

Eddie, "It means, well, in the future I won't be harvesting fifty acres of Christmas trees each year. I think I will be growing great nurse trees for oaks and maples. My little Wolf Christmas trees will be a thing of the past. I can see down the road very quickly. The idea of grass lawns, single wolf trees, and real Christmas trees will be stories for our grandchildren."

Julie, "What are nurse trees Eddie? You're not walking me down a dead end, are you?"

Eddie, "Nurse trees, soft woods like poplar, are the first trees that usually grow in the open, and they love and thrive in the direct sunlight. They nurse along the hardwoods that like shade to grow up in. I mean up. Eventually the oak, maple, and beech trees grow tall enough to shade out and kill their nurses. Not to worry, as the nurse trees eventually fall to the ground, rot and provide food for the trees that blocked their sun. Landowners who know what they're doing turn those nurse trees into a cash crop. Pulp and paper companies need poplar logs to make paper. Don't ask me too much, there is a big bunch of learning coming our way.

There are a number of trained arborists in the area. My friend Art has all of his woodlots under the Ontario Managed Forest Program. He'll give us lots of leads.

The good thing is that we've been feeding the squirrels, chipmunks, and the flying squirrels. It's been hard, but believe it or not, I've trained every one of them to gather nuts."

They all laughed and Eddie just smiled. If they had not laughed, the yarn would have continued. Training the squirrels to work with the chipmunks, then how to sort the wormy nuts from the good nuts, had the uneducated in awe. Eddie knew sometimes the truth is too true. He had often watched in the early fall as the red squirrels would climb a large oak tree, and would strip the acorns off, dropping them to the ground. They selected the best. Eddie watched one day as the chipmunks on the ground picked the acorns up, stuffed their stretchy skin cheeks, and carted them off. After a few minutes the red squirrel would come down, the chipmunks went into hiding, and big red would not find any nuts. He would go back up and toss down more. Squirrels have planted more nut trees than any person or tree planting company will ever come close to. They don't always bury the nuts where you want them. During the first or second year the seedlings are easy to transplant out of the flower or vegetable garden.

Eddie, "You must have some ideas on how to start this project."

Barbara, "We have a rough draft, but it is rough. Remember that all our students who went to Ottawa will be talking to their parents and friends. We will have to move quickly and get the ball rolling or as Jack would say get the spade digging."

Barbara, "Step one we need to talk to Tom and Veronica and get them on board.

Step two we should organize students to put ribbons on the trees indicating the water level in the trees. We need clipboards for the other grades so they can start their calculations."

Two hours later, Brent, Barbara's husband, arrived home. After the Lions meeting he had picked up coffee and blueberry scones for the team.

Brent "Looks like a serious scam going on in here."

Brent, "Eddie, how are your bees?"

Everyone laughed as they remembered Eddie's last fundraiser for the construction of the school's first horseshoe pitch.

Eddie had sold honey to all the staff, and anyone that would listen, for a dollar a pound.

Everyone knew Eddie was scamming them again but it would be worth the five or ten bucks to see what it was this time. A dollar a pound was way too low for honey. They would buy into the scam anyway. They knew the funds would all go to another worthy project. Eddie said they could only buy five pounds at a time and people understood and handed over the five dollar bill. Eddie would pull out a huge roll of bills, take off the elastic, and roll their five on the outside. Then he told them that he had only one bee, but it was working hard. Eddie explained that once the students finished the new horseshoe pits at school they could use the pits free while they waited for their honey. The donors always got something for their money.

Brent is an environmental consultant and a long-time buddy of Eddie's. They had gone through school and university together and were both hired by a large mechanical engineering firm. Eddie lasted only one project. His lack of patience was going to kill him, so he had to find another vocation and decided to go back to university and train to be a teacher.

Brent still remembers the day in Pickering at the nuclear plant when he tried to talk Eddie out of resigning.

Eddie had told him, "Brent, I have to get out. Look, we spent a year building that floating barge that we could flood with water. It sank perfectly and we were able to connect the elbow on the discharge pipe. That barge is a work of art and it works perfectly. There is no reason in the world that we have to build another new barge to put the next part on. I know our company is making money, I am making money, and you're making money, and there's lots of overtime, but what a waste of taxpayer's money and my time. No, I can't do it. In thirty-five years when I retire I will not be proud of myself. The number of people just trying to look busy..... no, no, and no. Brent, life's too short. I am out of here."

The rest is history. Eddie found his satisfying niche in life and Brent ended up as an environmental engineering consultant in his own company. Many of the assignments were for the provincial and federal governments and a big rubber stamp with the correct signature was all that was required. Brent loved to be schmoozing and making contacts.

On the redevelopment of the Toronto waterfront Brent was completely at home. The project involved four levels of government, three or four federal departments, numerous provincial departments, and a revolving door of changing city departments. Representatives of each kept changing with elections, promotions, and internal politics. It was a merry-go-round that paid well with lots of overtime and deadlines to meet to push the correct paper. Brent estimated that for every dollar that actually was spent on real goods like a walkway, retaining wall, stairs, or anything physical in nature, there were ninety-nine dollars spent on soft costs. These included salaries for permanent and temporary employees, meetings, studies, lawyers, and engineers. Most important, independent

consulting firms like Brent's company soaked up the 99 percent. Civil servants hate to make decisions so this is where
Brent's company came in to fill the vacuum.

One day when Brent and Eddie were trout fishing on the
Ouse River, Eddie asked Brent how he had become so successful and why his consulting firm grew so fast.

Eddie, " I met Trish Tucker and my business turned on
a dime. At the end of a meeting where I was involved in rezoning residential to commercial property, Trish approached
me after the session. She started asking questions. Trish
was writing an article for the local newspaper. I had time
before my next session, so I decided to explain what we did
as simply as possible. That's when my fortune changed.
Trish was a former high school English teacher who packed
it in to try writing, journalism, and photography. She told
me, "Anything would be easier than teaching. She did not
want to answer to a bell every day. Her dad worked at
General Motors in Oshawa and he grew to hate the bell
before Trish. In the next issue of the weekly local paper,
there I was, great picture and bold heading, **"The Expert
On Rezoning."** The phone started to ring off the hook.

I took out her card and called her up and asked her if she
would meet for a business lunch. We met and she told me
how she worked. Many magazines, newspapers, periodicals,
and other print materials pay to have someone write articles.
She was an avid reader and could take someone else's article
and turn it around and submit the same information to other
magazines. She had the ability to write at a grade five level

or any level you wanted. This paid the rent but the groceries came from the people and companies she championed and promoted.

My business took a jump just from one well-placed article. I signed on and my name, picture, and my company started to appear in all sorts of print media.

Many newspapers and flyers are looking for print to fill holes at press time. Trish had contacts just about everywhere and she had a file drawer of filler material. It covered hundreds of topics and on a minute's notice these articles could be faxed or dropped off in time to make the press runs and fill the empty holes. Today it is much easier to move material to meet deadlines. With Trish and her team the articles were elastic. They could be shortened or lengthened within minutes.

Many times I would hear something interesting going on and I would feed the leads to Trish.

At first I paid half the cost of ads as they were sourced for free by Trish, but this soon changed.

I was once a good church-goer and tithed myself to the church, but after all the situations with our church leaders we've drifted apart. I just tithed myself to Trish. She was shocked when I told her that I was paying one tenth of our gross sales to her company. Trish was able to hire two more part-time assistants and our business boomed. You get what you pay for; it's as simple as that.

There are many firms out there that are as good and maybe better, but we are top of mind and we get the first call."

Eddie: "How do you end up getting so many contracts?"

Brent: "I compete mainly against two other firms that are high profile like mine. The government requires three independent tenders. Eddie, I own, through silent partners, the other two firms. I make sure that the bids are very different."

Brent worked the system and was often called to make introductions and smooth relations, not only between the same level of government but in the same department. The money was very good. There were no bad debts to try and collect.

Brent's ski chalet in Bromont, Quebec, was always stocked full of food and booze when government employees arrived with their family for their holiday. The correct number of lift tickets were on the dining room table. Brent's second consulting firm owned a fishing lodge, once again in Quebec, on a private well-stocked lake. If a department or division contracted over $200,000 a year in fees, the head was invited for a four day fishing vacation of a lifetime. The third consulting firm had golf memberships at three or four exclusive clubs and that brought in more business.

Brent did not tell Eddie about the perks he received back from the engineering firms that won the government contracts. After all, Brent was married to Eddie's first wife's cousin. Eddie would not understand. Brent never paid a cent for the ski chalet, fishing lodge, or any golf membership. He just passed the perks along the line and kept the contracts and fees rolling in. Brent understood the rules of the game and played them well.

Trish became an expert at writing the same environmental news in different ways for different audiences and free copy was easy to sell. Brent kept the leads flowing.

$\backsim\!\!\!\!\!\backsim$

Eddie, "Brent, we need your help in funding a new project at school."

Brent, "Barb told me that at lunch this week, you pulled one of your best pranks. I have to hear about it first."

Eddie said, "Oh yeah," and broke into uncontrolled laughter, tears coming to his eyes. He was speechless for a change.

Eddie slowly regained control and looked at Brent in the eyes. "You know, I can tell you all about the gag and you'll get a big bang out of it." Eddie looked at his watch. "I have to go to the half-ton first as I have to put some water in the cage."

Brent, "What have you got in the cage?"

Eddie knew he had him hooked. Could he reel him in?

Eddie looked at Brent with a straight, serious, and concerned face. I have live trapped a cross-breed no it's a hybrid. I think it's part weasel and part mink. I caught it near the old rat farm on the Mill Line Road, north of Norwood at the broken dam. Its head is weasel-like and the body is the size of a full-grown mink. The funniest thing you have ever seen, it's white, and this time of year it should be dark brown or black. I've had it about three weeks and I'm running out of frogs and mice. It sure eats a ton. I call it Wink, short for weasel and mink. I thought of calling her Measles but that sounds like a disease, so Wink it is. There's no question that it's a female. She is pretty for a member of the weasel family. You know, the Ministry of Natural Resources heard about it and they want to come and pick it up. I don't trust those guys. I'm not so sure they're going to get Wink. I'm thinking of letting her go back to the marsh where she came from. Today the biology classes at the high school had a good look and they took a number of pictures."

Barbara and Julie hid their eyes under the table as they knew the trap was being put into place.

Julie spoke up first, "Eddie can we go out and see it?"

Eddie, "Not a chance. If it ever got away I'd be in trouble. I could bring it in here and if we're careful, you could look into the peep-holes in the box."

Eddie had borrowed the gag box from his brother. Alley was a travelling plumbing and heating salesman, and it was in Marmora or Madoc that the bar owner had used the gag to fill his tavern full every lunch time. The locals would gather for a beer and hot sandwich every time the blind was pulled half-way down the front window. It was the signal. They knew another person would be sitting on the stool peering into the peep-holes. Two beers, a hot sandwich, and a good laugh were hard to pass up.

Eddie looked at Barbara, "Would it be okay if I bring in the box cage?"

Barbara, with all her acting ability, slowly said, "I guess it would be okay."

Eddie had to decide who was going to look into the holes first. Brent was a pretty big guy to catch.

Eddie: "Julie maybe you should be the first to see Wink." Eddie knew Julie had seen this gag before and she would be part of the team.

Julie knew what was coming and she had to bow out without tipping her hand. "I am very leery of animals. I'm not sure you should even bring it into the house."

Eddie: "It can't get out. The lid is locked."

Brent bit the bait, "Eddie, go get it, I'll have a look."

Eddie: "Brent, you sit here at the end of the table. Barbara and Julie, you sit behind and please don't move when I bring in the box. If you talk just whisper, as we don't want to frighten Wink."

Eddie was soon back from the half-ton and the shape of the box was visible under a dark blue car blanket. Touching the blanket, Eddie said, "If you keep the light out they're much more quiet. The blanket does the trick.

Brent, I am going to slowly take the blanket off and then I'll let you know when you can look into the holes at the front."

A small piece of balsa wood was at Eddie's end of the box, and as he took the blanket off, he scratched the wood with his nails. There was no question the animal was in the box. You could smell it.

Brent bent down and looked carefully through the holes into the box. He saw nothing. "Eddie I can't see a thing."

Eddie scratched the balsa wood again.

After a couple of tries, Brent was still unable to see a thing in the box.

Eddie looked at Barbara, "Do you mind if I just open the lid a bit for Brent to see Wink?"

Barbara: "Just make sure it doesn't jump out. The kids are upstairs asleep."

Eddie: "Brent, get real close and I will just lift the lid a crack. Then you can see Wink."

Brent peered forward and Eddie tiped the lid off the box as he flicked on the light switch at the back of the box.

The stuffed black squirrel with lights glowing for eyes and large white teeth, shot out of the box. The three-foot spring projected it over Brent's head and he yelled and fell backwards into the arms of the two waiting ladies.

The white teeth were donated by a trapper. The small squirrel jaw was stretched and spread to get the large incisors in.

After Brent recovered from the initial shock he could not stop laughing. The laughing, not the yelling, woke the children and they came down to the kitchen to see what was going on.

Brent would agree to find the funding for the School Tree Project.

Brent, "I have a number of contacts. They have money sitting in their budget that they have to flush. I'm doing them a favour. He smiled. "They will need a submission, and don't worry, I have a template and Barbara can help with a few key words. The tax receipts I can pick up at the school board office. Karl, the director, is retiring and I've found a couple of corporate boards that need him to fill out their mandates. Good pocket change for Karl and something to fill in his time. As to the new director's position, well, Trish and I have our person well-positioned in the queue. Don't worry about media coverage. Trish will get the project lots of ink.

Eddie, "You work in a different world than the one I know."

Brent, "You join a team and you play for life."

Eddie was one of those teachers that everyone connects with. The new grade eight students walking into Eddie's class the first day in September were always surprised. One year he had two canoes hanging from the ceiling with a sign-up sheet to paddle down the Ouse River to Rice Lake. Another year, when the thirteen and fourteen year old students walked into his classroom, the first thing they saw was a large tent pitched at the back of the room and four desks in it. The sign read, "We are camping out a week this Thursday, get packed." In a number of subtle ways everyone was invited into his world. Eddie took the time to know the name of each student before they came through the door. He had picked out his volunteer class monitors before they arrived. Eddie hated with a passion, filling out permission forms, collecting for this and that, and all the other paperwork that bogs a teacher down. The neatest, most particular student ends up with the fun responsibility. A monitor to take attendance, a monitor to take stuff to the office and other classes, a monitor to post the day's date and weather, a monitor to list the scores of all the teams playing in the area every season-Eddie was a master delegator. Grade eight students liked working and Eddie kept them busy.

Eddie cycled to work every day he could and would often embarrass the parents of his charges. In the morning many parents would be in their SUVs, idling at the side of the road with their children waiting for the bus. They would sit behind the wheel with a coffee in one hand and a munchie in the other. Eddy would have five different routes, one for each day of the week. He would take turns riding out and saying "good morning" to his students as they waited for the yellow and black bus to arrive. At first parents asked their son or daughter who that stranger was that said hello. The usual answer was, "Mom, that's Mr. Patton, my teacher. He cycles to work and often times goes out of his way for a bit more exercise."

At the first parents' night, Eddie would have his notes and would score his points. He would slide the "fat mobile" in at the appropriate time. "When I ride out and see parents sitting with their children in the family SUV, well, I just nickname them fat mobiles."

Eddie had a lunch-time cycle club, and had a system in place. Students that did not have, or could not afford, a decent cycle were looked after. The local hardware store was owned by one of Eddie's former students. Most of the staff also were former students. A small town is like that.

To ride well, you have to have a bike that fits. Older brothers and sisters would pass down their wheels. Some of his charges were the first in the family line and there were no older brothers or sisters to pass down a bicycle, let alone clothing. If money was a problem, Eddie had a system to get them a part-time job, or a trade situation to get the cycle to join the group. The one or two students each year who did not have a cycle were introduced to Clyde, the owner of the local hardware uptown. A trade deal was always struck, pre-engineered by Eddie. Eddie often said that was why he studied to become an engineer.

How the deal worked was the young student would pick out a bike from the catalogue that they thought they might afford. A trade was struck. Sometimes it was twenty chickens or three turkeys. The non-farm kids often shovelled off the snow on a warehouse, or walked a route delivering sales flyers to prospective customers. Eddie insisted that it was not a hand out but a hand up. (This is the motto of Habitat for Humanity) When the day arrived, usually a week later, the bike was delivered, things changed. What the young thirteen- or fourteen-year-old student ordered never came. Instead, a much better cycle arrived and in perfect proportion to the young rider. Having a cycle that fits you makes it so much easier and so much more fun. Eddie had sponsors that would chip in. The hardware store charged their wholesale price, less ten percent, and the funds were found outside. Eddie kept a mental note and he was not surprised, the students that were helped now were often the adults down the road that pitched in the most.

Eddie insisted that every cycle came with a locking system. It was safe at school but you had to be aware.

When Dougiee got home from Ottawa he ran over to see Mr. Bob, they had a great time as Dougiee's stories helped Mr. Bob re-live his own trips to Ottawa. Mr. Bob asked a thousand questions and got a million answers. Mr. Bob said that the first couple of years there would be a shortage of seedlings and they had better get collecting seeds from prime trees.

Mr. Bob: "Dougiee, don't forget to tell your friends to collect seeds from a number of trees, not just one. Just like breeding the stallion, Walter. You want to hedge your bets. Those trees are going to be growing a long time and you don't know what future diseases are going to be coming through. We lost most of our horse chestnut trees, just about all of our elms, and now the ash trees are dropping like flies. Plan on a variety of trees, not just the ones you like. Collect seeds only from the best, healthiest trees."

Dougiee's parents forgot or didn't realize their son had been away.

On Sunday, after Dougiee's short time with his parents he headed over to Mr. Bob's. Once the chores were done they went into the tool shed and pulled up two stools. Mr. Bob had cleared off the big wooden work bench. The clutter of bits and parts had been removed and the thick wooden surface looked like it had been dusted. The binders of breeding records were stacked in the corner. Hanging on the wall were three large sheets of white newsprint. Graph paper, pens, pencils, and markers were stacked in a blue Player's tobacco can. Mr. Bob did not have to tell Dougiee that this project was important.

Mr. Bob had never planted an acre of seeds so he just worked it out. If each seed was three inches apart, there could be four seeds per foot. Two hundred feet would mean eight hundred seeds per row. If the rows were a foot apart you could have one hundred rows in an acre so eighty thousand seedlings per acres sounded about right. "Dougiee if you planted three small trees abreast, three feet apart, along the highway, you will need 5,280 trees a mile just to plant one side. If the property owner on each side of the highway did the same you would need 21,120 seedlings per mile. One acre of seedlings will only plant about four miles of roadway. We need to collect seed and get a bunch of farmers to volunteer a few acres. I'm going to talk to Eddie as he's grown his own seeds for years. He plants about fifty acres of Christmas trees a year, so he must have made all the mistakes by now."

Mr. Bob, "Three feet apart is too close for full grown trees. When they get bigger, you can prune out the weakest and any other trees that show bad signs for the future. Just like planting carrot and radish seeds. You can crop two or three times then let the rest grow." Mr. Bob had an analogy for everything he told Dougiee.

Dougiee, "We have no other choice. Half my class come from farms so you can count on them getting their parents to help out. Mr. Bob why not get the government involved?"

Mr Bob, "My experience is they would have to study it, form a committee, hire consultants, put it out to tender, argue if it is a provincial, municipal, or federal issue, no matter in a couple of years they would download it to the lowest level of government. This is too important an issue to let the four layers of government get involved. We can spend our own money better than letting them spend our tax dollars. Dougiee it costs the government twenty-five cents to collect a tax dollar and it costs them another twenty-five cents to spend a tax dollar. We've lost half of our tax dollars before they buy anything." Mr. Bob rarely did so but he turned his back and spit.

Mr. Bob took a pause and then continued; "Ontario had one of the best tree seedling and tree planting departments in the world. It was a model for other countries to follow. Guess what, the Premier just about closed it all down and most of the trained professionals vanished along with their expertise. There are only a handful of private companies that have now filled the void left by the provincial government. You also have to realize that all three provincial political parties have been in power since then, and they all have done absolutely nothing to restore the department, but good talkers and rule followers.

The Ontario government has maintained a tree seeding facility but it won't be large enough to handle the demand for seeds. Private groups will have to step in and supply the seedlings. There is practically no cost, just time and patience.

Private organizations like the Ontario Woodlot Association could do the job.

In the United States the two universities that could lead the way are Oregon and Cornell. They have an enviable record, the resources to get the word out about re-planting our forests, and the contacts to turn government legislation around.

Dougiee, the United States is so rich and bountiful that they subsidize farmers or land owners not to grow crops. They produce too much food and it lowers the market price. To keep the supply in line with demand they take farms out of production. Imagine land that is sitting fallow could be growing seedlings and/or trees, and helping lower the ocean levels. Every country has to work together on this project."

CHAPTER SIX

Letters to Jack Wilson

J ack arrived home one day from work to find his wife, Joy, in the kitchen preparing Jack's all-time favourite meal-macaroni and cheese. Maybe it should be called cheese with a bit of macaroni thrown in.

Joy, "Over the last three days when you were away, a pile of letters arrived for you. From the writing on the envelopes it looks like they're from students. Most of the postmarks are from Norwood. There's also a large brown envelope from Julie and I think it's full of envelopes."

Jack could hardly wait to open the letters and start reading them.

"You know, I've promised all the grade six students that I would write them back if they wrote me on their progress. After I've read each letter, I'll write a short note back. Would you mind checking my letters, as you know my spelling and grammar isn't that good?"

Joy, "Sure, I'd really like to read what they have to say. I feel so out of touch with young people today. When Chip was growing up we knew what was going on with his age group. It seems like so long ago."

That evening, Jack and Joy sat down and opened up the envelopes and read the student letters.

Jack had made it a practice to write a note to an employee or customer every day he went to work. He went out of his way to find something positive to say and would take the ten minutes to hand-write a note to drop in the mail, or put into the employee's mailbox. Few people today ever receive a hand-written note telling them that they are a valued customer or employee. Sometimes it would just be a note saying, "Chip, I noticed today that when I got home you had shovelled the sidewalk and driveway. What a treat to come home and see it all done. Thanks Son." Or "Sara, thanks once again for making all the arrangements for the western swing through Vancouver and San Francisco. Every detail of the trip worked out as planned. It's reassuring to know that you're so particular. I can relax knowing that you're running the office when I am away. Thanks Jack."

Jack pulled out his grandfather's treasured pocket knife and cleanly slit open the first envelope.

Dear Mr. Jack:

My parents don't believe me that trees are full of water. They think I am making it up. They saw the news last night on CHEX TV when they were showing forest fires out of control in Northern Alberta and California. They said water does not burn. I told them your story about the water in popcorn boiling then exploding and Dad laughed at me. He does believe that being in the shade is cool. Two years ago the two large bamagillia trees on the sunny side of our house had to be cut down as they were starting to lean towards us. Bamagillia is a name of a poplar tree that grows in our area. Last summer my parents could not sleep upstairs in their bedroom as it was too hot. They slept downstairs and Dad has let a number of the root suckers grow that are popping up next to the two big stumps. That is called coppicing I think. We are planting trees for shade not for water storage all over our small property.

Susan Baskin and I play on the same softball team and our Dads sit together and talk. I think that is why dad is helping me plant seeds.

Mr. Jack I tried.

Yours tully,

Susan Martin

Ps I weighed popcorn before and after I popped it and Dad said if I did not stop making popcorn, the neighbours will think we are opening a movie house. Popcorn does not have much water. The steam will burn you.

Dear Susan,

Thanks for taking the time to write. Some people need to see evidence first hand before they believe. Try getting some fresh cut small pieces of wood. If you can get the pieces the same day that the tree is cut it will show the complete effect. Water starts to evaporate the minute the wood is exposed to air. If you can get pieces around one or two kilograms that would be the best size for kitchen scales. Put up a small sign, just a piece of paper folded like a sandwich board. Each week weigh the pieces and record the results. During the winter when the humidity in your home is lower the drop in weight is faster. Convincing a doubter makes for a strong advocate. When they tell other people, their word of mouth is much stronger. Mr. Bob would say, "When the jackass from Missouri speaks up, people listen." "I am not saying your dad is from Missouri, but you know what I mean."

Let me know how it goes.

Jack Wilson

Jack; "What do you think of that response?"

Joy: "I am not sure that you should call Susan's dad a jackass. You've often said blood is thicker than water. You can get the same support message across another way. Mr. Bob would say, "When a team of horses are pulling a load, they can pull a bigger load if they work together."

Joy continued, "Here is a way you could get your Dad to pull with you." Planting trees for shade is a reason strong enough on its own. Let other people and time convince your Dad that those trees are full of water. Be patient. Trees are here for a very long time."

Jack: "You know Mr. Bob?"

Joy, "I feel like I know him as you've mentioned him and Dougiee so many times."

❧

Dear Mr. Jack Wilson

Mr. Jack I am in grade eight and we have not heard you speak or watch you draw. We feel left out and I asked my teacher if I could write you and ask if you could include our class next time when you come. I checked and my brother in grade six said there is room at the back for our class. Our teacher Miss Barbara Brown would allow us to go.

Lorraine Parker

Ps. Next year I hope to go to Norwood District High and I want to start a tree planting club. Dad has already planted pine and spruce seeds for next year. Ten buckets full of seeds took all of my friends all day to plant.

Dear Lorraine,

Let your teacher know that its okay with me if your class attends. I'd also be happy to come and visit your tree planting club next year when you're in high school.

Jack Wilson

〰

Dear Mr. Jack:

My dad is related to Billy Gallagher that used to own the sawmill on the Ninth Line north of Norwood. He knows a lot about trees. He laughs when people say the original forest. It was not the original forest when our ancestors first came here from Ireland. The original forest would be millions of years old and long dead and decayed into fertilizer for the next growth. Dad says people should say, "The first cut." Trees were at their maximum size and water holding capacity. Dad says leaving your tall trees makes the new younger trees grow tall and straight. They grow up to get their heads above the others to catch the sun. Each generation of trees grows taller, each trying to get their heads above the others. Dad says it will take a minimum of four generations of trees to get back even close to the height of the first cut. In our area white pine grew over two hundred feet tall that is seventy meters or more and the base of the trees were two meters across. Some pines were three meters or close to 9 feet in diameter. We had a picnic at the park in Haliburton. The old steam powered sawmill is in bits and pieces but you can tell logs were very large back then. Mr. Jack why did they cut all the tall trees?

Mom and Dad say they did not know any better back then, but we are still doing the same.

Sincerely

Bobbie Bennet

~

Dear Mr. Willson

We had to write a letter to you.

Thank you for visiting our class.

You draw on the board really well. I am practicing and my trees are starting to look okay. I guess.

My sister are and I the only two girls on the worm picking crew. Our ankle cups that we wear to pick worms hold 10 seedlings each. Every night we go out we plant 40 trees. People in town are used to seeing our cap lights at night, but they don't know we are planting trees before we start picking.

The guys made fun of us planting trees. They were silly and we bet them.

We over picked our hot spots enough each night to win. Everyone all six of us are planting trees every night. Herb the boss found out what we were doing and he is supplying us with willow shoots. He says the wet areas were we pick is ideal for willows. The two golf courses have every corner and fence line planted. We keep them out of the mower path.

My sister likes your hat, Rink Rats. Could we have one, I would like one, how much will we have to pay? We have the money, worms pay well.

Yours truly,

Wanda Williams

Ps This is the first letter I have ever written.

Mr. Wilson my sister and I have never received a letter. Could you write us and mail it from one of those far away places, like British Columbia? Is that in Canada?

Dear Wanda,

You will find enclosed two hats for you and your sister. Don't tell anyone how you got them. Just to let you know, provinces have funny names. Newfoundland and New Brunswick are not new. Prince Edward Island does not have a prince living on it, but there are a few princesses if you know what I mean? Women live in Manitoba. British Columbia is not British anymore, it could be called New China. Wanda, many Chinese families can trace their ancestry back to the building of the Trans-Canada Railroad. The British think they are the important ones and Columbia is not the same as the southern country. British Columbia is the province to the far west and it is drop-dead beautiful. Chip, my son, will be there next week on Salt Springs Island and he will mail this letter for me.

Think about this: I put snow tires on my truck so I can go ice fishing.

Keep on planting. Every tree counts. Put them back.

Jack Wilson

Joy: "I don't think it's a good idea to give the two girls free hats. What are you going to do when the other kids find out? Besides, people have a sense of pride. They're working and earning money and they want to pay their own way. Think of your work and how you get paid."

Jack: "Never really thought too much about it."

Joy: "Your cousin Bruce, fishes and he has lots of friends who fish. Why not barter? You could negotiate so many dozen worms per hat. They would like that idea. Your cousin would be surprised to get free bait and just might make the connection to set up his friends to get bait from the girls. I think you could put that scenario in place with a few phone calls and a drop-in visit to your cousin. We haven't

seen Ruthie or Bruce for a long time. Why not invite them
for a visit or we could drop down and go fishing with them
for a day?"

Dear J. Wilson

"Mr. Wilson my cousin in Norwood has teased me for
years that Mount Forest has no Mountain, now he has added,
we have no Forest. Dad and I will plant 50,000 seeds down
the side of our back field and we talked our neighbours into
matching our challenge." Guy Van Schwindt in Norwood
will plant 30,000.

We want to put the rain drop on our big black birch tree
but Dad thinks it is dying.

Sweder Van Schwindt Grade 5 Mount Forest

Dear Sweder,

I am happy to hear you are planting seeds and getting
started to put the forest back into Mount Forest. Let the big
birch tree wear the blue ribbon until it dies. You can move it
to another tree when the time comes.

Jack Wilson

Mr. Jack Wilson

I told my parents all about you. They grew up on the farm next to yours.

We found out how to gather white pine seeds. You have to get ahead of the squirrels to get the good ones.

Dad says there is a pine called jack pine but he says white pine grows better in the areas we are planting. My parents are having a seed planting weekend and we hope to plant all our fence lines. We are planting three abreast and on the road side we are planting on both sides of the fence like you suggested.

Will you be visited our class again?

Wendy Webster

Dear Wendy,

If you watch and you are careful, the squirrels will gather the nuts or cones for you.

Next spring the little trees that sprout up can be transferred to an open area.

Jack Wilson

Dear Mr. Wilson

We put the rain drop on the tree down the street. It is very old and very large, lots of water. People don't believe us when we tell them how much water is stored in a tree. Mom says people don't believe what they can't see.

Robert Webster

Dear Robert,

The steady rain soaks. The weigh scales study is the only way to convince some people how much water is being stored in trees. Just to remind you, when the sap is running out of the maples in the spring, many people still do not understand that water is stored in the trees all four seasons.

Try the weigh scales demonstration. Cut small green branches from a maple, oak, or birch. Then cut the stems into pieces, six to eight inches long, like asparagus. Put an elastic around them and weigh them daily to see how fast and how much water is in the branches. You could also use cedar branches. Good luck.

Jack Wilson

Dear Mr. Jack:

I am winning the bet with my dad and older brother. He said that trees do not hold 50 % water so we have a bet. I really wanted a new snow board for next winter.

One of our sugar maple trees blew over in a strong wind and we cut it up for firewood as it was one of those wolf trees that you drew on the board. My older brother Jeff laughed when I called it a wolf tree. They think I am making things up again.

One of the big branches was fresh and green, not like the center of the stump. The chunk was about 20 centimetres in diameter and 30 centimetres long. Mom says I have to put wax paper down on the scales when we weigh it each day. It has been four weeks since we started to weigh the block and my brother is getting scared. He will have to pick my share of stones this year off the back field. Dad says I should start to pick out the board.

Mom says sales of garlic and horseradish has been good this winter and we will be taking a ski trip to Whistler. I told them the storey of how you fell off the Canadian Olympic team and they want to try the Saddle. Mom wants to ski down the run where Nancy Green won.

My lacrosse coach says that the farm kids that have picked stones with a hay fork are the best throwers so I might have to trade with Jeff.

Dad has been making so many fishing baits this year that he has hired his best friend Joe Longfoot to help out. They laugh when they work and they don't let us into the shop as the baits are real secrets until they are sold. Lots of secrets when you go fishing. Dad says Mom is a great bait designer and gives her all the credit for the baits doing so well.

Dad and mom have bought another farm and he said he would plant half of it in seedlings if we can find enough seed.

Can you let us know where we can buy seed. White pine, and spruce. We have lots of maple seeds.

I bet Mom and Dad about how fast ice melts and how the water sublimates. I explained sublimates and I think I got it right. Solid to vapour with no bath in between, I think that is what you said.

My ice block is under the shield I built and theirs is out in the sun. My drip pan is filling up and theirs is half as much. Dad thinks I have a future career with the gaming commission of Ontario.

When are you coming back to visit our class? Dad says you would be my third cousin. He explained how it all worked but I got lost on the cousins.

Susan Baskin

Joy, "Susan is Ruthie and Bruce's daughter. She would be a third or fourth cousin."

Dear Susan,

I hope you will tell your story to as many people as will listen. Enjoy Whistler and Blackcomb. I hope to make it to Le Massif and Mount St. Anne this year. Mention to Bruce that the Ontario Woodlot Association would be able to help him find tree seeds for your area. It is important to grow the trees that are from your area, as their success rate will be far superior to trees out of your climate zone.

Your third or fourth cousin,

Jack Wilson

❧

Jack also received a letter from a parent, Eldrid Quakenbush

Dear Jack Wilson:

You may not know me but our grandparents farmed close to each other north of Norwood. Your Percheron horses were from our stallion, Big Dan. Dad said your father had the best boar in the area and our Miss Bell and Miss Tinkle would have large litters every year. Over a dozen piglets at a time.

I just want to tell you that our son, Percy, has for the first time been excited about school. He has calculated the water content on just about all of our trees within sight of the house. The big spruce is wearing the rain drop. He wants us to calculate the water in the back bush and it is over 20 acres. I am not sure how to do it. Percy said, "Dad it is one tree at a time." He says, we have to put them back.

We dug over 200 small seedlings out of the woods already and put them down our laneway three abreast five feet apart just like you suggested. The row in the center will pull the wolf trees up on the outside. We have planted seeds along each of our working fields and in the corners where we can't plant crops. Percy has all the window sills and all the walkways filled with spruce seeds. The barn and drive shed are surrounded with growing seeds.

Percy is doing his homework every night after chores and he says he wants to go to tree college when he graduates from high school. Looks like he will have to go to Quebec as Ontario has closed just about everything down. Percy has taken a real interest in French and we can't help him much with that topic. He is talking to the seeds and trees in French.

Percy looks forward to getting on the bus to go to school. We have one of the largest wood lots in the area and under the Ontario Managed Forest plan it is in very good shape.

Percy has gained new stature as he knows all about the three basics. Ground cover, stems and canopy management. He has memorized all the different tree names. The class has been out twice to walk in our woods. The logging trails are pretty rough but they don't seem to mind. I think they just like riding the sleigh or wagon behind the Percherons.

Thanks for catching our son's attention.

Eldrid

Dear Eldrid,

Dad told me that Sammy, our boar, had a smile on his face every time he came home from your farm.

Hope we can get together when I am in town next time. Lots to talk about.

Jack Wilson

          ⟳

Dear Mr. Wilson

I was able to get a scale in Madoc as we were going to visit Uncle Swift in Smith Falls. Every store has sold out their scales and has ordered more. I have been able to get fresh pieces of birch, pine, maple, and ash and you are right. They lose weight fast as the water evaporates. We put the rain drop on an old elm in the back yard. Dad and Mom said not to get my hopes up as the disease is still around and so far it has missed ours.

I am growing horse chestnut seeds as I like the big white flowers in the spring and we gather the nuts in the fall. Dad uses the shells to make his famous antique dies. It takes more room to plant horse chestnuts so in the small garden we have and along our driveway I am growing white oaks.

My dad thinks I have gone nuts. My little brother Frederick is helping me.

Our teacher calculated that our class alone would be planting over a million trees next year.

Sincerely,

Bruce Burgess

Dear Bruce,

Keep planting. More people should be a little nuts.

Jack Wilson

⁓

Dear Mr. Jack.

We live in a rented apartment above the hardware store. We don't own any property so we don't own any trees. My best friend Charlie comes to school on a bus and he lives on a dairy farm. I ride my bike out to his farm and we get to play after the chores are done. We are moving cedar trees out of the woods and planting them along the farm pond. The farm pond is fenced so the cattle can't pee and drink at the same time. This is a new law and it is working. The Ouse River runs through their farm and the cattle are not allowed to drink out of it. Cattle are funny, they bite, chew, swallow and step forward until it is time to chew their cud. Cattle pee

245

and poop and do not stop eating while they do it. Cattle eat small trees and Charlie's mom has put up an electric fence so they can't eat the trees we are planting. Farming is fun but the work never ends. I put my rain drop on a big soft maple by the Ouse. It is the one that has an old hay fork rope hanging from a tall branch. We swing on the rope all summer and drop into a deep part of the Ouse. It is not the biggest tree but we all love it.

Someday I want to own a bit of land with my own trees.

Yours turly

Cliff Cuthbertson.

Dear Cliff,

Work hard and plan for the future. There are lots of city students that would love to have a friend like Charlie. Enjoy what you have and thanks for sharing your story. A good friend, a strong rope, and a good tree make for a fun day in the Ouse River.

Jack Wilson

Dear Mr. Wilson:

My brother Doug and his buddies got into trouble last year planting marijuana seeds along the side of the road. They got the seeds from an old package of canary seed they found at a garage sale. Mrs. Metcalf had really old stuff when she died and it was all for sale. Someone squealed and the police arrived at our house. My parents said they would put him in the dog house if he didn't straighten out. This year the boys planted squash and pumpkin seeds all over town. Doug has been nicknamed Seed by his buddies.

They said they would help me with the tree project if I would just stop bugging them. Susan gave me four large sap buckets of maple seeds and we have made spike poles to stick in the ground so we can plant the seeds. Mom gave us a shaker of cayenne pepper to put on the ground after we plant the seeds to keep the squirrels away. My rain drop is hanging on my neighbour's maple tree. He was surprised when I told him how much water his tree was storing. Mr. Calendar is really old. He is over sixty and I am not sure he understands that if the tree does not store the water it would end up in the ocean. Mr. Calendar has never travelled outside of Ontario or Quebec and has never seen a real ocean.

Bye

Wesley Thomas

Ps My brother and his friend got the pumpkin seeds out of Mr. Brown's garden.

He wins the contest each year for the largest pumpkin and squash at the Norwood fair and he will not sell or share any seeds. Doug and John one night smashed open one of his pumpkins and a squash and took out most of the seeds and left a dead ground hog beside the pumpkin. They stuffed pumpkin in the ground hogs mouth. John saved the seeds in a wooden basket and they planted them this spring all over.

Pumpkins and squash are growing all over town this year. Don't tell anyone I told you.

Dear Wesley,

Thanks for helping Mr. Calendar understand about the water in his tree. He will tell his friends and they will tell their friends.

Jack Wilson

⌒❦

Dear Mr. Jack Wilson:

As president of the school council I was hoping you will accept our invitation to speak to our parent's group. If you would be willing we would arrange to have a parent's meeting the evening next time you are in town. This could happen after you are at the school talking to the grade six classes.

Time is money and I am not sure how much you charge to speak.

Please let us know if you can fit the parents in to your busy schedule.

Yours truly

Bergen Tedford.

Dear Bergen,

There is no charge for me to talk to the parents' group. I would be happy to talk to the parents' group and any other adults who wish to come.

Jack Wilson

Mr. Wilson:

I have proposed a special project for our 4 H club. We are going to call it "Hay it's raining." We are going to link all the clubs together to monitor the evaporation challenge you gave us. We have members in just about every county in Ontario. Some counties do not have any farmers. We are going to gather those things called statistics and see if Hay it's raining can be shown to be true. We will be starting it soon. I am in charge of our county and I first have to list all the farms and how many working hectares they have. Next I have to determine the number for each crop and pasture. We will be ready to start recording next spring. My friend Alfreda is responsible to get the other Ontario county 4 H clubs on board. Alfreda can talk anyone into doing anything. She convinced our class to go on the bridge over the narrows and jump in after our graduation. She had it organized and only two kids had to be pushed a little and Sammy was shoved. Everyone had fun even Sammy. We are going to record the crops planted and date and follow them through to harvest. We are going to record the weather every day and see if there is a pattern. Our leader says we may have a chance to get other provinces and even the States involved in "Hay Its Raining."

It sure shocked our club when I told them that farmers create the weather, especially rain. I explained that hay is 95 % water and when millions of acres or hectares are cut the water moisture goes straight up and forms clouds then rain for the farmers down wind. If farmers worked and planted together they could harvest before the rain then have the rain give a boost to the second crop. We get a lot of heat over genetically modified food. Wait until they hear about farmer controlled rain.

All the members in our club are growing tree seeds. We have all ordered rain drops and will be putting them on our largest trees close to the house. Every member has made a commitment to plant a minimum one acre of seeds. You can plant over 60,000 seeds in an acre.

Mr. Wilson, farmers feed cities and we also can help shade the planet.

Alfreda is going to contact Rural Youth Europe and All China Youth Federation and see if they can start their own "Hay it's raining" project.

Alfreda made a bunch of friends at the last international conference. I didn't have enough money to go.

She is also going to tell them about the Norwood rain drop project.

Sincerely,

Samantha Brown

Dear Samatha,

Wow, you have taken on a huge project but you also have a huge organization to work with. The 4-H club commands respect and they have earned it because they have people like you committed to a farming way of life. I am very interested in your rain project. Please keep me informed.

Jack Wilson

❦

Mr. J. Wilson;

I have declared war on all lawns. We should plant more trees and the shade should stop grass from growing. We have to put the trees back and I put my rain drop on Dad's John Deere riding lawn mower. He said he would get rid of our back lawn first if I come up with a plan so it doesn't look like a waste land. I then put the rain drop on my neighbours riding lawn mower and he talked to my dad and they both have agreed to turn the back lawns into a forest. The rain drop is now on the big white pine in our side yard.

Mr. Wilson you are right parents and grownups will listen to you.

Two bad kids down the block were making fun of me yesterday and said the rain drop was a joke. My older brother heard what was going on and he told them if the rain drop went missing he would hold them responsible. Mr. Wilson I am not allowed to swear, he really said he would kick their ass black and blue and he would talk to Uncle Bill, their hockey coach and they would not make the team this year. They got silent real quick like.

Sincerelly,

Pauline Stackhouse

Dear Pauline,

Sounds like your big brother is a real supporter. Glad you have started the war against lawns.

Jack Wilson

Dear Mr. Wilson:

We are busy planting the roadside beside our farm. Dad has made up a tune. The nearest coffee stop is 16 kilometres away and our ditches are floating in their cups and plastic caps.

Roll up the rim to win.

Come back tomorrow and Try agin.

An empty cup I don't need

Down goes the window

Out goes one more Timble weed

Just Blowing in the wind.

Dad says they should start a campaign, roll up the rim to win and if you lose we have a tree for you to choose. Every cup would be a winner and the customers would have a tree to plant. Spruce, cedar, maple, and oak, all the local varieties would be available. Millions of trees could be planted.

How would we get in touch with that building called Head Office? Dad says the local owner who sponsors our hockey team would help out. Dad also said that the head office makes up the rules.

Yours Truly

Dick Clysdale

Dear Dick,

Thank you for the little ditty. Maybe there should be a contest for the best ditty.

Jack Wilson

CHAPTER SEVEN

Mort and Molly (The Water Molecules)

Barbara, "Jack, my students understand the food pyramid and the life cycle, but I'm not sure if they have caught the concept of trippage."

Jack had an idea.

When Jack was visiting the grade six classes in Norwood he mentioned to the students that he would like them to write a letter to him and pretend that they are Molly or Mortimer, a water molecule. The idea was that Molly or Mortimer would move a minimum of three times. "This is not a test. I just want to see if you understand what I mean when I use the word trippage."

Jack assumed the girls in the class would pick Molly and the boys for sure would pick Mortimer.

Without knowing all the ears were listening, Jack said; "If you write a letter to me I will write back." But then he dug himself even deeper. "It's been a long time since I grew up in your town. If you have time, write and tell me what you like to do or what your family does for a living."

Were Jack and Barbara ever in for a surprise!

Barbara was pleased that every student wrote a letter to Jack. Many students usually found it painful to write, but not this time. The students wanted to talk to Jack and this was their way. After Barbara had read all the letters she decided not to red-ink the corrections. She thought Jack would enjoy the uncut versions. She would photocopy the letters, correct the spelling and grammar, and hand the photocopies back to the students. Dougiee wrote two letters. One she would have Dougiee read to the class and the other letter was just for Jack.

Barbara had each student read their own letter to the class. When the students returned from afternoon recess they had time for two students to read aloud. Not only did it settle down the class, but it gave students oral practice in reading.

Dear Mr. Wilson

It was fun working on the blackboard. I think you and Mr. Bob must be related. He lets me do a lot of stuff. Mr. Bob teaches me and he listens very well. You listen with your eyes just like Mr. Bob.

I have cousins in Toronto and they live on streets where they have signs up that you cannot play ball hockey. The school yards are fenced in and they have taken down the basketball hoops. You are not allowed to skateboard on the streets. Mr. Wilson they sit at home and play games on their computers and get fat. Your off side comments are funny but true. I tell them about walking four or five miles after school and shooting ground hogs and they think I am out of the dark ages. They are busy shooting down air planes, and killing thousands of people on their computers. My two pet hens lays eggs and I show them how I wash off the chicken shit and they get upset. They do not know where the eggs come out. I tried to explain how eggs are candled and they grossed out on me. Mr. Wilson, city people do not know where their food comes from. They will be in trouble in our next disaster. They will not survive when the grocery store runs out of food.

The Purple Heart is taken so why not a rain drop? Most of my classmates are busy playing games and reading books about witches and goblins. Planting seeds and growing trees is fun and we can see things grow. I think I can get a group of friends to help.

Trippage

I looked it up and there is no such word. You made it up didn't you? You are like Mr. Bob. If he needs something he makes it.

Mortimer was a drop of water on the glass of water sitting in my fathers' den last Sunday when I had a chance to talk to my parents. It was picked up on my mother's hand when she reached for a cool drink. She wiped her hand on her napkin and it was carried to her room after our meeting. Mortimer evaporated and was carried out of the house through the bathroom ceiling fan. Mortimer, lighter than air rose up with all the moisture being released by the crop of barley next door. The winds were pushing the clouds east. Dust from the thrashing in Foxborough rose high into the sky and provided the resting place for Mortimer. Rain fell on Smith Falls and Mary a patient at the home saw a leaf with a water droplet. A small bird stuck out its tongue and licked off the drop of water.

Yours truly,

Douglas MacArthur (Dougiee)

Dear Dougiee,

Thanks for helping out when I talk to your class. Mr. Bob is my dad's uncle so he would be my great uncle Bob. Sometimes life gets busy and you do not stay in touch. I would like to drop in and have a visit with your Mr. Bob and Grace next time I'm in Norwood.

Jack Wilson

Dear Mr. Wilson

Mortimer fell from the sky into the back yard and he was frozsen and with other flakes his own age and sise. He had travelled from the surface of Goregian Bay and finally decided to fall out of the sky. He was just submated up in the air. In the spring he warmed up and turned into water and found an old worm hole and he and his buddies fell down into the bottom of the round cave. A small root sucked him up into a bigger root then finally he was lifted up above ground in a trunk of a large maple. We were all carying sugar and heading up the tree when all of a sudden we hit a slide and fell out of the tree into a sap bucket. We sloshed around in the bucket for almost a day and then we were dumped into a bigger pail by a young girl named Megan and put in a big tank with lots of our friends and many we did not now. We were all carrying sugar and having a good time. It was no time and we were in a tube heading for a big hot tub. The firewood underneath was burning and very hot and many of my friends dropped their packs of sugar and went out the roof as vapour. Katie was condensed in the heat exchanger above the boilin sap and she was fell down the stainless steal pipe into a stainless steel tank. Peter produces Maple water. Maple water is the purest water there is. Peter explained to the school group that was visiting. The 60 to 150 year old maple trees filter the ground water, and the boiling process of the maple sap separates the water from the sugar and other minerals.

I saw Peter and he was busy explaining.

Peter points to the condenser, "And this stainless steel hood collects the heat from the steam coming off the boiling sap. The heated pipes then pre heat the cold sap coming into the boiling pan. The steam condenses on the pipes and this pure maple water flows down this stainless steel pipe into the stainless steel container. All the stainless steel is food grade stainless."

I hung on for dear life and it got realy hot and it was boilling until finally someone turned on a tap and we were dumped threw a filter. This was the third filter but not the last. We were put into a clean plastic pail. It got really cold. We waited a week or too then we were placed in a small hot tub and boy did we get hot again. Some of my friends, like Katie were caught off guard, lost their grip and ended up flying away as vapour. Once the heat was turned off and we cooled down we were put into clean 750 ml bottles and a label was put on the front. "Sugar House Maple Syrup"

Oops Peter is pouring us into a pot on the cook stove in the back room and I think we are going to be taffy. I had better hang on tight as more of us will be vaporised. Daryl, my grandmother just came in the door. She was driving the team and wagon and showing all the other people the sugar bush operation. Not to many of my friends have a grandmother that is teaching them to snowboard and how to ride a horse.

Whew, made it and look out I am falling into fresh white snow in a trough beside the sap house. Look out there is a popsicle stick heading towards us and oh no He has braces on his teeth. He is kind of cute. I may be saved, I am stuck on the braces and fighting for my life. The mouth has closed and the salliva gun has washed me off and down I go with a few friends.

This is very warm and there are acids all around me eating all the lumps.

I think we are stopping at Leaskdale fries stand. His dad says, "They are the best." His mom always says, "You say that every time we stop at a country stand. Mom and I think the Fry Guy in Anten Mills is the best. I have a friend, Carol and she has five older brothers and when she is talking to anyone of them she always says, "You are my favourite older brother."

In the washroom and down we go to the septic tank then out into the weepers. It is dark and the odour is, like it just stinks, Mr. Jack.

I seep through the layers of sand and gravel and into the ground water. The sand and gravel scrubs me clean again.

Mortimer, I mean Mort, was sucked up by a cedar tree on the fence row next door to the chip truck. Torture time. You smell the food but you are trapped in a tree. As Mortimer was being drawn up the outside of the tree just under the bark he met a number of molocules that he had seen in Georgian Bay years ago. Mr. Jack water molocules don't age and they have been here a long time so they have many friends. In the cedar tree near the heart wood Mortimer saw his friends, Sinclair Green and Flora Sink. They were resting. Mr. Jack is that where you got the name Green Water Sink? They mentioned that they had been trapped in the heart wood for almost 100 years and they felt that they would soon be moving on.

Each year people are spreading fertilizer and weed kill chemickals. It is starting to smell like chemickals in here.

It had been a good rest but they could hear the black carpenter ants heading their way. Have you ever stood beside a herd of cows eating dry hay? The sound of ants in a tree is real noisy. They were eaten by the carpenter ants and sure enough moments later a pileated wood pecker drilled through and boy do they have a long quick tongue. They were down the beak in a flash. Sinclair and Flora were regurgitated, that's a bit like vomoite into the beaks of two young ones. A few hours later they were pooped in the bottom of the nest. They soon would evaporate and be on the move again. The smell was terrible.

I ended up in a cedar seed.

All summer I watched tradesmen and smelled locals and tourists eating French fries just beside me. It was torture Mr. Wilson.

In the fall when it was starting to cool down a Cedar wax wing ate me. Thats a pretty bird.

The ride was short. Sitting on a hydro line just outside of Leaskdale I was squirted out and fell with the seed into the ditch. In the fall, then the winter covered in snow and ice and as soon as spring warmed up the seed started to grow. The road salt just about killed the seed. I hung on to the salt until the sun freed me. I left the salt behind.

Dad calls the cedar tree the cafeteria tree. Beaver chew on it, deer eat the cedar browse as high as they can reach. Squirrels and chipmunks feed on the seeds if they can get ahead of the birds. Our neighbours cut the cedar browse and sell it to the oil makers and we sell cedar posts to the wine industry and we also saw the logs for decks and siding. Our living room has a cedar ceiling and it smells nice.

While we were eating our French Fries dad explained how they take a 5 or 10 cent potatoe, push it through a set of knives drop it into a boiling fat of oil and presto a bit of catsup, vinegar and salt and you have taken 5 cents and turned it into $ 5.00. Mom quickly looked at dad. "You should complain," Dad works in the beer industry, "You work for an industry that takes a liquid worth less than 10 cents put it in a returnable bottle and a case of 24 sells for over $ 30.00."

My dad has a great sense of humour. He reached over and took mom's chips and dumped them into his and picked up her cardboard container. Mom just stared with her mouth open. Dad stood up with his poker face and said, "I am going over to see if there is a deposit on this container."

"I usually win the family groaner of the day but dad beat me this time."

If we plant more trees we would all be better off. My friends tell me I never stop talking.

Yours

Megan and Katie Neve

When will you be back to visit?

Dear Megan and Katie,

You certainly have an interesting life and you do understand Trippage. Your teacher will let you know when she has time for me to come back and visit. Maybe at sugar maple time we will meet again.

Jack Wilson

❧

Jack decided to read just one more letter before going to bed...so he thought.

Dear Mr. Wilson

I think I understand what trippage is.

It is spring and there are lots of mosquitoes and one bit me and sucked out a bit of blood and I missed it and it got away. I go the next one. Our teacher says at our age we are 65 % water so Mortimer was on his way. That evening a small frog stuck out its tongue and ate the mosquito. The frog didn't know enough to be quiete and started crocking and a blue heron at sunset had a snacks. Dad often says I talk too much but no one eats me.

The blue herons are the only bird my dad says that when they take off and let go can lay a 50 foot chalk line.

Mortimer and the whole chalk line was dumped into Stoney Lake, just down the lake from our cottage. I think it is called the Duck Pond. Mortimer was taken by the slow current and two months later was washed over the small dam at Gilchrest Bay into White lake and then down to Indian River and eventually he would have ended up in Rice Lake. A deer running across the River in Warsaw picked up Mortimer on her chest hairs and carried Mortimer into the woods. Mortimer was brushed off on a small birch tree.

Mr. Wilson this could go on for ever....I get it, that is what is suppose to happen.

My friend Larry Parker came over and he said my skate board was calling and I have to go. I need to put on new bunks and wheels and Larry is as handy as a pocket in a shirt.

No wax.

Bill Scott

Thanks Bill, for your letter. You do understand trippage.

Jack Wilson

Jack thought, just one more letter;

Mr. Jack:

Mom says I have to write tonight or I can't go to the skateboard park.

Mortimer dripped off the roof of our house and landed on the grass. The grass sucked up the water and was cut by the lawn mower. Mortimer evaporated with all the other Mortimers and went way up in the sky. Clouds formed and it rained down east on a skateboard park. A dog licked the puddle and Mortimer was later peed on a fire hydrant.

Good bye.

Cecil James

Dear Cecil:

Yes you know what Trippage means and you are fast.

Jack Wilson

Mr. Jack Wilson:

Molly was in a snowball I threw at Butch. It was a good shot and the snow went down his neck. Molly went home with him and was absorbed in his shirt. He took Molly home. I live in a group home on Dummer Road and I have never had parents. Molly has a ride and will be in their home for some time. Molly evaporated out of the shirt and is floating in the air. She has been breathed in and out by Butches parents all their children and even the family dog, Shirley.

I want to be in a home so Molly will be staying there I should say here for awhile.

Mr. Wilson do you have brothers and sisters, you have not talked about them? What are they like? I am not sure what and ant and an uncle are. Old people are grandparents. Do they act like they do on TV and Christmas? I have never been taken to a hockey game or a ball park.

You have kind eyes. Will you write back?

Sinserely,

Jenny Cooper

PS This is my third set of foster parents and they are really nice. I am behaving myself now. I am starting to like school. I hope my worker does not come and take me somewhere else.

Next time I have an interview with my big sister from Keene I will try to keep my mouth shut.

ᘓᕂ

Hi Jenny,

You do understand what Trippage is all about. You get to pick your friends but you do not get to choose your family. Sometimes families get along but most don't. When people get older, they usually spend more time with their friends than their family. Big sisters can be better than family. Have fun.

Jack Wilson

ᘓᕂ

Mr Wilson:

Thanks for visiting our class. I have tried to draw like you and I am getting better. You only say good things and I like that. Most people are complaining or arguing and when I go home I make a snack and go to my room to read.

Molly and Mortimer are holding hands and they go over the falls at the dam. They are on the surface and are evaporated and float over to the hay field near the Crowley dairy barn going to Hastings. The apple tree has ripe apples and the evening condensation settles. Eileen picks the apples and takes them to school for her friends. Will Scott, he is cute, is given one of the apples and before he eats it he rubs it on his shirt. Molly and Mortimer are on the school bus back to the farm. Will gets home and changes his clothes and heads to the barn to do his chores. He is in charge of all the new calves and once they are pail fed he works with his sister to bring the hay down out of the mow to feed the waiting mothers. Young calves have to learn how to drink and you get a pail of milk and when you put your fingers up their mouth you lower their head into the pail. You can really have fun. When town cousins come out to visit you will set them up. Filling the pail too full or letting the calf drop their head too low in the milk pail is a set up. The calf will inhale the milk and then lift its head and blows out. Get out of the way!! Lots of fun on the farm.

Molly and Mortimer overheard Will's parents talking about the cruise they were planning to go on. When they found out that the cruise ship dumped all the black water and kitchen waste back into the ocean they decided not to travel on a floating toilet. They checked out the other cruise ships and found that all ocean going vessels are floating toilets. Military, commercial and cruise ships all dump toilet waste into the ocean. They have decided to go to Irwin Inn for a week and golf the local golf courses. Those big ships carry 8,000 people and they are dumping into the ocean every day.

I checked with my cousin who lives in Halifax and he says all the cities around the world that are on the sea coast dump raw sewage into the oceans and seas. Seven billion people dumping in the oceans are too many.

Mr. Wilson when whales beach themselves are they telling us they don't want to swim and live in our shit? Dad says I can't use the word shit. He say I should say defecate. Do people know what defecate means?

Yours

George Granger

Dear George:

You definitively know what Trippage means. It is sad to say that there are only a few ships that don't dump their black water and waste into the ocean. You are also correct that over two thirds of the world's population lives along the coasts and they continue to flush their black water into the seas. Seven billion people every day and more people coming. The recovery rate for the oceans and seas will soon be tested. If we do not change I think we are in big trouble.

Jack Wilson

༺༻

Mr. Wilson,

Molly was in the strawberries shipped to our grocery store from California. Mom and dad have an artesian well on our property and he is not allowed to bottle the water and sell it back in Ireland. There is a law that we can't ship water out of Canada. Dad and Mom say all the fruit and vegetalbles we import from the States and overseas is 85 % or more water.

I told my dad that farmers help create the weather and he listened to me.

When you cut 50 acres of hay and you want it to dry, when the field of oats matures and turns brown, when the fields of wheat turn brown, when the millions of hectares of corn mature and turn brown where does the water moisture go?

I told dad that the mosture like it is water vapour it goes up as it is litter than air. He laughed at me. I then said ice is heavy and it floates on water. Water vapour is light and it gets pushed up and is cooled when it gets higher. Those clouds you see are farmer clouds up wind. Our house is up wind of the manure pile and up wind is important. My favourite Aunt Margaret sells real estatge and she says if you are on the west side of a city property values are higher. If you are selling commuters like to be on the east as they do not want to drive into the sun going to work or coming home. Smell is impotant on a farm.

Coniferous trees hang onto their leaves called needles for two years. They would be nude if they dropped everything. Pine, spruce and fir make up a big part of this group. There is a rotation system going on. Like a bingo dance in the gym.

New growth this year hangs on over the winter to dry up and fall off the next fall. Tamarack thats where my dad plays golf. They drops all their new needles every year. These needles are dry and make an acid bed on the forest floor. This preventing completion from others.

Deciduous trees look like, maple, oak, birches, ash, aspen drop their dry leaves every fall and stand nude all winter. They sure look cold when the snow drops on them.

They let go the water, boreal forests and farm crops cause the fall rain showers. Mom is funny and my friends all like her. She hangs out the wash every Monday on our cloesline. She is also causing rain down wind.

Morning mist rising off the river, lake or ocean is the visible time to witness the amount of vapour constantly lifting off the surface of exposed water. My cousin calls the morning mist sea smoke.

Missy Jackson

Dear Missy:

Thank you for the letter. Living on the farm you get to see many things first hand.

Jack Wilson.

Mr. Willson:

Molly fell on a clover leaf in the field next to our home. Bulla the Holstein next door ate the clover. Bulla can make the loudest gas of any cow you can imagine. Molly was chewed with the first cud and went into the second stomack and was made into milk. Out of the teet down the long house into the stainless steel cooling tank. Next morning the tanker backed in and swoosh Molly was in a tanker heading for Plainville. The hoses were connected and Molly was dumped into the cheese factory holding tank. Wayne Lain the head cheese maker turned the valve and Molly was heading to the pasteurizer. Opps Wayne smelled and tasted the milk and decided instantly that this milk would by pass the pasteurizer and head to the vat. Old cheese is made from the best milk and it is not pasteurized. Molly was on her way.

Molly went trough the process and was heated, cooked, cut and put through the curd mill. Molly liked the press the best as she hung on when many of her friends were squirted out into the whey and they would be going to the centrifudge and then maybe off to Sterling to make Whey butter. The others were headed to the pig farm. Molly stayed over night in the press and next day she was in a block and headed for the cold storage room. Molly would be trapped in the block for over eight years. Old cheddar sells for a premium price because it is so good and Molly would be resting for awhile until someone ate her. Molly was not in the piece of old cheddar that I brought to school and gave to my teacher. Molly likes my teacher.

Linda Skinkle

❦

Dear Linda,

Thanks for your letter. You understand trippage and you are also correct about cheese. I often stop at Maple Dale and buy their eight year old cheddar. It is the best next to the fresh cheddar curd. When it squeaks you know it's fresh.

Jack Wilson

❦

Hi Mr. Wilson:

Trippage is being on the move and changing resting locations. Sometimes you are sucked up a weed, eaten by an animal, boiled, frozen, smashed, or you could just be floating up from a fresh cut field of hay.

Floyd Stockdale

Dear Floyd:

You sure know what trippage is all about. Thanks for your letter.

Jack Wilson

∽

Hi Mr. Wilson:

(Paul did not write a letter he just made a sketch.)

The sketch clearly showed:

The water moisture condensed on a leaf of a milkweed pod dripped off onto a frog. A green grass snake opening its mouth to eat the frog, from above the outstretched claws of a hawk were just about to seize the snake.

Under the picture Paul wrote, "Lunch or be lunched?"

Sincerely,

Paul Watkins

∽

Dear Mr. Wilson

Molly and Mortimer were sitting on a stool

You will see neither are a fool.

Molly was eating curd and Mortimer was too

The whey dripped and stuck to their fingers like glue

One lick and it was gone

You should hear their love song

Evaporation took its toll and they were off and away.

You can follow them to Trent River by the Bay.

Morning mist like sea smoke, they awoke

Rising in the sun they looked at each but did not spoke

Speaking is better than spoke but the frog, no, down the throat

Splash the bass under the lily pad

Mr. Wilson I am sad.

I understand trippage but you can talk and draw pictures for everyone.

Sinserelly,

Bernice Williamson

Dear Bernice,

Thank you for what you have said.

All is not lost if we look ahead.

One step at a time, we will restore

Nature and be able to live next door.

Jack Wilson

PS You have a way with words.

∾

Mr. Wilson:

Jim Fife took a pass from Bob Delaney and he saw Bob Webster heading up the left wing and put the puck on his stick. Six, seven strides and a head fake and the puck and Mortimer were in the net. The referee picked up the puck and Mortimer and rubbed him off onto his glove. After the game the glove and Mortimer froze in the grip in the back seat. Tim Chapman the referee was dating his wife to be and on the way home from the Square dance in the car and contents took a side lane for some fun. The heat of the moment released Mortimer and when the windows were opened for fresh air Mortimer was on his way east.

Dad works at the car wash and he told mom that storey. I was listening from the living room.

Mortimer got pushed up into the air and became part of a storm cloud to fall in Montreal. Henry Richard was getting out of his car and Mortimer in a snow flake fell on Henry's toque. Got to tell you the fun of walking into the arena and down the aisle into the bar set up for retired players. Henry took his toque and shook it all over his buddies. Mortimer was hoping to land on one of the great players of the team, like Boom Boom, but landed on the floor. He was later swept away and deposited in the garbage bin next to some old French fries and half eaten hot dogs. Mortimer stayed frozen in the dump over the winter and later seeped into the ground and leaked into the St. Lawrence River. Mortimer realized that this was the closest he would come to fame for years.

Mortimer was swalled and chewed by many fish until one day he was on the surface and the sun cooked him. He was vaporized and he was off to China again. He didn't like the pollution but what could he do, he was being carried by the wind.

Sincerely,

Peter Walsh

∽

Dear Mr. Jack Wilson

I live on a farm east of Norwood and we pass the half eaten Esker every day. It does look ugly. Will the tall hydro towers topple? My brother, Dave, wants to go to university and he is smart. He mentioned to dad and mom that if he raised rabbits he could get enough money ahead so he could continue after high school. As long as he does his regular chores he can raise rabbits. They multiply fast and he has a large market in Toronto and Montreal. I helped to build more pens and have learned to clean out the pens quickly. I don't get paid. My brother says this is all mine once he leaves home for university. I am working hard at school as I know I will have the money to go.

Molly was in the corn cob that was throwed into the rabbit pen. Miss Bee was ready for shipping and Molly was off to Montreal. In the pen on the way Molly was deposited and evaporated into the air around Kingston. The rain storm moved in and the droplets falling caught Molly and it landed on the roof of Sir John A. MacDonald's home. As the water was falling off the roof it splashed and was caught on the boot of a tourist and was carried into the house. Molly fell off in the library and was absorbed into the carpet. Mr. Wilson, I love history, Molly will be staying in the library for a long time.

Sinserely

Arness Boate

Ps My brother gets all the throw outs at the local vegetable and fruit stand and he has fixed a number of deep freezers that were being throw out by our neighbours. He does not have to buy much feed for his herd and his customers all tell him his rabbits taste the best. We have to be very careful cleaning out the pens and taking the manure away from the pens. All the predators smell the manure and they are looking for the rabbits. Dave traps them and he sells the skins for extra pocket money. Our uncle has a trapping licence and he sells the skins for Dave. I didn't like skinning and scrapping but I have to realized that if I want to go to university, some jobs need to be done. We work together and I get half. By bank account is growing. Dave is teaching me all the Italian and Greek names I need to know to sell the rabbits. Oh, one other thing, do you know there is no fat on a rabbit and if you only ate rabbit you would starve to death? Is that right? Dave wants to be an accountant. He likes numbers and is always using his calculator to figure out costs and weights and all those things he thinks are important.

Bye again

Arness Boate

Mr. Wison:

My dad is in the concrete business not cement business. Cement, gravel and water is mixed together to make concrete. Dad says it is a chemical reaction. I just know it is messy and it gets hard.

Molly was poured into the big truck and the load of concrete was on its way to a bridge on highway 401 in Toronto. Yonge street is the longest street in Ontario and where it crosses under the 401 they seem to be always expanding that concrete work. Within two day as the concrete was drying Molly was released to move on. The wind pushed north to

Lake of Bays and Molly fell out of the sky hanging onto a dust particle from Sudbury. Their big stack is very tall and dust gets spread further away. The drop of water hung under the leaf of a morning glory and a Black fly came along and we were gone inside the fly. The fly didn't make it very far as a mosquito hawk scooped it out of the air. A phoebe ate the mosquito hawk and Molly was deposited on a clothes line and fell on a white towel. Boy, birds and clothes lines are a problem. Molly evaporated but left the stain behind. The air currents pushed her south and down she came in a storm in Norwood. Molly landed on our roof and made her way down the pipes into the cistern. Dad has a growing business installing concrete cisterns. Most of the new homes are getting cisterns. Mom did the laundry and Molly held onto a cute blouse and Molly is in the closet watching me write this letter.

I hope that explains trippage.

Bev Wood

Dear Mr. Jack Willsen:

I sure liked the shoes you wear wearing when yous visited our clas. I have three older brothers and four older sisters and I am the babby of the family. I try really hard at school but my marks are low. I get lots of help at home but book work is not fun for me. Dad taught me how to weld with the electric welder and the propane oxyeyen one and he says my stuff sticks better than he has ever seen before. I dirve all the equipment big and small. The big John Deere has been adjusted so I can reach the pedals and the corn cuter has may favourite CDs. Every year I spot the incexts before everyone else and I have won the ploughing match twice.

Mr. Willsen I understand trippage, Molly got into the potatoe chips and they were too be throughed out but I put them on a cookie sheet and heated them up in the oven. They chips were fine. Molly left the oven and stuck to the ceiling of our kitchen. Once the door was open she flew the coop. The cloud was blowed down wind and Molly fell with all her friends on a horse in Fernleigh. I want a hors and mom and dad said after the soybeans are in and we are paid they will pay half. Got to go Iam the goalie on our team.

Dougiee is the only boy in our class that is kind to me. He nows i work hard and just doesn't get it like my older sisters and brothers. They all are on the honer roll.

Yours trully

Angel Wheeler

Dear Mr. Wilson:

Mr. Bob and I look after Dougiee. He is a gift to us both. He talks about you non stop and we are trying to catch on. Mr. Bob is turning his back fields into nurseries for tree seedlings. When school tours come out I bake cookies and treats for them. Young people have a good appetite and it is easy to feed hungry people. I go to the seniors club once a week and I was wondering if you could come and speak to us. The members have families and they could spread the word to plant trees. The seniors club would like to get involved and they have been around the block a few times and they have the time.

Yours truly,

Grace Bedore

❧

Hi Mr. Wilson:

My dad always uses annalegies to explain things. He says our planet is like a big pine log. The carpenter ants are in it and they are chewing it apart to make homes and trails. The pine beetle is not looking for a permanent home. Pine is what they want to eat. Dad says that is what we do. We chew up the surface of earth, dig holes and make homes to cover up good farmland and forests. He says the tar sands, gravel pits, mines most things we do are like ants eating up and destroying the planet. The pine beetles and ants are all working away but they don't realise that together they are going to eat them selves out of a house and homes. Ants and beetles can move to another log. There are lots of logs. We have only one planet. Dad says humans don't deserve planet earth.

Mom likes the word trippage and she said I should start Mortimer in the ground. We are farmers and ground water is important too us.

Mortimer started in a snow flake the sun came out. He got hot in the sun, melted and fell to the ground. Mortimer was picked off by a pine fiver root. He was molecule up into the tree trunk and was in the growth layer for nine months then it happened. The storm rolled in and the freezing rain got so heavy it snapped the top right off the big pine. Dad and my two older sisters cut the pine tree up for firewood and the four bottom logs were pulleds to one side of the barn yard. The Kubota has a wnch on the back and it can pull logs out of the woods. Dad had a plan. Russell Thompson came by in the spring after half loads were over and he brought his portable sawmill. It is an orange Woodmiser. Dad decided to leave one log on the ground and he would saw it a year later after the worms and ants had put lots of holes in it. He wanted to use the holly lumber for furniture.

Mortimer is in the sap wood on the bottom of the log. The frozen log thawed out in the spring sunshine and the Pine Beetles found the logs on the ground. You could hear them coming. They shewed holes through the bark and into the soft white layer below. Three female Black carpenter ants found the log on the ground. They had just had fun with their partners and they were looking for a home to lay eggs. Jack, the males die and the ladies get to go on. Mortimer did not know if he was going to be eaten by a pine beetle or an ant. He was lucky this time. The main ant tunnel went beside him. A paper thin cellulose wall was the curtain between the action and Mortimer. He watched all the dead ants each night being hauled out. Then it happened. He hear a diesel engine and then he felt the log rolling over and moving. It landed hard and it jared him and vibrated through the log. A few hours later he felt the log rolling and then the log was still. He cold hear the gasoline engine of the saw and he started to hear the blade enter the end of the log. The blade passed over head a few times getting closer and closer. The log was then turned 90 degrees and the saw sound passed over two more times. They next time the blade cut so close. The board was being lifted and moved. The board dropped and Mortimer was on the underside of the board and then evaporated and was gone down wind. In a cloud over Baie Saint Paul he fell on top of the gondola at le Massif.

I have too go. My sisters need me to shoot the squirrels that are trying to chew through the boards into the grainery. Mom knows how to cook them in the crock pot and they are very good. My friends at school think my sandwiches are chicken and I get to trade for lots of good stuff

Sincerely,

Winston Young

Dear Jack Wilson:

This is my first letter.

I live on a farm North of Norwood. The land is not very good for farming as it has too many stones and very thin topsoil. They say it is just like Ireland where my great, great grandparents came from. Ontario white cedar trees grow well. We only get one hay crop a year while farms south of us can grow two or three crops of hay a year. Dad did not need to use the equipment very much so he started to work for other farmers. He said the business just grew busier each year. He kept buying more and larger machines and soon my older sister and next my two older brothers were all needed to work with Dad. Our family is in the custom farming business. My mom and dad, two older brothers and my sister are never home. They call me the afterthought. I am much younger and it is just me at home in the spring and fall.

My parents and my brothers and sister move south and north following the planting season and as soon as it is over they are heading south with the harvesting equipment. In the winter everyone is home. We have a large work shop. The outdoor furnace heats all the water for the radiant floors and air handlers in the house. The work shop radiant floors are very comfortable when you are on a dolly under a machine changing a drive, a chain or just changing the oil and filters. Dad replaces parts before they wear out. His Uncle, my great Uncle Greg, was an aircraft mechanic at Trenton. Dad says parts wear out and it is much cheaper and easier to replace them in a warm shop with all the tools than lying on your back in a grain field. That is called preventive maintenance and Dad says that is why the business is growing. Fewer break downs, lower cost and on time customers are happy.

Got to go to bed.

Dad wants me to go into the business. Dad put me in charge of our parts room. I order and stock all the parts that wear out each year. Dad was very surprised when I computerized the parts inventory and ordering process. I also have a spread sheet for each piece of equipment and they are used to me bugging them when they call home every Wednesday and Saturday night. I record the hours or kilometers and remind them of all the filters and fluids that need changing. Sometimes they don't want to stop and do maintenance. They know they will pay the price later so they get out the portable lights and work into the night.

Dad wants me to go to college and learn all about robotics. He believes farm machinery will soon be operated out of a control room and the machines will not have an operator on board.

We square bale our hay. Most of our customers in the States prefer large square bales. Not the round ones. Pierce, my oldest brother, he is 27 years old. He always brings things home for me. A few years ago Pierce piled large square bales around the house after hay season. The bales are about 10 meters from the house and just about as high as the house. We live on top of a drumlin so it is very windy and cold in the winter. It looked funny and many people made comments. Mom really didn't like the idea, but she said, "Go ahead and give it a try." The first winter we reduced our wood consumption by one third. Our neighbours are starting to pile their round bales to break the wind.

Trippage is not a word. I have checked for it everywhere. I know what it means. The sod behind the big John Deere turned over in Kentucky, dried and Molly was released and was pushed high into the sky. The fine ash from the coal fired generators stuck to Molly's new work clothes and she fell on my dad's new float. They were all heading to Virginia to plant peanuts in two days. Molly got a sun stroke but lifted off the float and was pushed west as a storm was coming in and blowing everything out of the way. Molly stuck to a leaf on a plum tree and the smell of flowers was very pleasant. A honey bee rested for a second and Molly was on his hind leg. Talk about a busy bee. Molly was carried back to the hive and was brushed off near the comb where they were feeding royal jelly to a larva to make another queen. It was dark inside then a large bang happened. The hives were being loaded onto a flat bed and being moved.

Mr. Wilson they move flat beds of bees around the country following the nectar harvest.

Molly ended up in a hot honey room and the machine was separating the honey from the frames. In the melting process Molly escaped and hung onto the ceiling until the door was opened. She was off again.

Got to go to bed.

In March we have a maple syrup operation and everyone is working hard, fixing lines, tapping and getting all the pumps working and the tanks cleaned. If spring is late my brother, Harry, stays behind to finish the run. I am in charge when Harry leaves. Last year I finished the syrup, put all the syrup in bottles and labelled them. A case of 12 is heavy to put up on a shelf.

Got to go to bed again.

Mom drives the highway coach. The big bus is three times as old as me and is completely rebuilt. It has been converted into a home and kitchen and it goes with the crew all spring, summer and fall. My Sister one day called it the Winterbago. We all call it the Summerbago. Each grain truck pulls a large camper trailer. Its a carvan. I get to sleep outside in a tent most of the time. There are some areas in the west that it is too risky and Dad makes me come inside.

Mom's older brother, Uncle Ned and Aunt Eva retired too early, Dad says. They sold all their property and live in Florida six months and then they look after our place for six months. Dad calls Florida, "God's waiting room." They come and stay with me and look after our farm. When school gets out they put me on the bus in Peterborough and I catch the plane in Toronto and join my family. This will be my second year. Mom can't wait to go each year. She grew up in the prairies before she married Dad and she is so happy to get back. She never stops smiling even in the wind and rain storms. After 10 days of constant wind and the dust has seeped in everywhere mom is still smiling. I know why my mom didn't want the house sounded by hay bales. Mom likes to look at the sky line far away. Mom looks after the meals, laundry and new customers. You should see how our bus is painted. It is a billboard. Every time we move to a new farm near a town, which is often. Mom parks the bus in town next to the farm supply store. We get new business this way. In the evening she pulls in with supper. We drag a small all wheel drive car behind the bus. The car has winter tires on it. When the water and sewage lines are connected mom uses the small car to run errands.

My Dad, sister and two brothers all have their AZ driver's licence. Each person drives a float with one of the big harvesters on it. Each year Dad hires three drivers for the grain trucks. Some of the guys come back year after year. Dad makes sure they like mom's cooking before they are hired as everyone eats from the same pot. This is a lonely job, you and the machine 10 to 12 hours a day. Day after day you have to like being by yourself. You are on a team but it is a lonely job.

Off to bed.

Last year Dad bought me a dirt bike so I can run errands. I don't like the name gopher so they all call me "Fetch." Mom and Dad call me April. We all have a walkie talkie and we are on the same channel. Dad reminds us that other people may be listening in, so we only talk when we have to and only business. Mom has two walkie talkies and she has on one the local channel wherever we go. Knowing the weather a few hours before it hits is important. At first I did not understand when they spoke. Mom is from the west and she understands. Mom alerts everyone to what is coming. I am busy all day long running between the machines and the bus where we have a spare parts bin. I update the laptop every night and check in with my friends at home. I deliver snacks and cold drinks out to the crew. The machines never stop unless they break down. The harvester unloads on the move into the grain trucks follow along beside. I ride in the grain trucks and I have a pole to pass out bags.

Dad makes everyone run with him for an hour after supper. Sitting in a machine all day even if they are floating seats is not good for the body. Dad also has installed rubber pull cords in all the cabins so you can exercise as you work. We bump into many farmers that are overweight. They see us running every night and they just sit around the BBQ with a beer and wave at the crazy Cunucks.

Last year was my first year and I guess Dad forgot to put me on his check list. I was delivering a price list to the next farm. Farms are really big out west and I keep my tank full and refill at lunch time. When I got back the bus and all the machinery had moved on. I looked at my walike talkie and the battery had died. Dad always told me not to panic if I was lost in a store or when we took a trip. "Stay where you are and someone will always come and pick you up." Well I was not lost and I knew they were moving three concessions up and one west to the next harvest. My tank was nearly full so I cut through the back roads. The floats had to go on the main roads. I arrived just after they noticed I was missing.

Dad gave me an extra hug that night and he even slowed up so I could run with him.

Bed time.

I fly home a week before school and the same bus picks me up at Pearson airport. The corn crop is getting harvested later and later each year so my family usually comes home in November some time.

Got to go, an order just came in and I need to pack up spare shear pins and courier them to dad.

I understand water trippage. When I explained it to mom she said we are a trippage family moving from place to place. By brother thought that was really funny.

Yours truly,

April Payne

Ps. Mom helped me with this letter. It took all week to write. I understand why gypsies don't like staying in one place.

∽

Dear Mr Wlsen

My older sister is in Mr. Patton's grade eight class. At home she calls him Eddie and one day in class she forgot. He just laughed. She won't do that again. She really likes Mr. Patton.

Mr. Patton gave them an assignment and I am helping my sister.

Mr. Patton explained to the class that a cruise ship has 5,000 passengers and 3,000 crew and they are sailing for 8 days. They flush their holding tanks out at see when they are sailing between ports at night. The cruise ships get a heavy fine if they flush in the harbours. A picture of a new very large cruise ship was pinned to the bulletin board. Mr. Patton pinned a large picture of a toilet beside it. Their math homework for the weekend was to sign up for one calculation and some could sign up for two or three if they wanted to.

Monique brought home the sheet and she had signed up to calculate toothpaste, shampoo, and makeup.

On the top of the sheet in brackets;

(8,000 people in 8 days consume and it is all flushed into the Ocean)

How many tubes of toothpaste are spit down the sink by 8,000 people in 8 days.?

How many bottles of shampoo are washed down the shower drain by 8,000 people in 8 days.?

How many tubes and jars of makeup are washed off?

Monique said that Mr. Patton asked Lucy to do the questions on drugs. Her mom and dad are both pharmacists and he knew they could help Lucy. He said to Lucy that it would probably take a couple of weeks and this would count towards one of her assignments.

8000 people, probably half women and half men, children, parents, grandparents. Good thing Lucy has parents working in a drug store. They probably know all the drugs and who uses them. Mr. Patton says the body does not retain the drugs for very long. They would be flushed into the holding tanks.

My friend Alwen picked the question about food. Alwen's parents manage the grocery store for Mr. and Mrs. Spencer. Mrs. Station is the cafeteria supervisor at the high school and she also is the assistant manager at the golf course in the summer. They would be able to help Alwen. Roland, Alwen's Dad, was trained at Camp Borden and understands what it takes to feed an army. All the left overs, carcasses, bones, used cooking oils and grease went into the holding tanks after the huge garburators chewed them into bit sized chunks.

Mr. Patton said to Alwen, 8,000 people eating for 8 days is a ton of food. Norwood has a population of 1,600.

Alwen's Dad had a tally on how many weekly shoppers and half shoppers and skimmers frequented the store. Alwen asked his dad what skimmers were. "Allen, they are the shoppers that just come in and load up on the specials, pay and walk out the door. We break even or lose on every one of those customers. They also lose. When new people move into town and they ask for recommendations on whom to hire, where to buy other products the skimmers are never on the list. Basil Williams the local plumber can never figure out why his competitor ends up with all the new customers. That's how small towns work."

Alwen already had figured out they would eat the complete store at least 5 times and drink the Havelock beer store three times. Alwen thought to himself, that is just one ship every 8 days and Mr. Patton said there were over 250 cruise ships and they are building more big ones.

Yours Isabel

❧

Dear Jack:

My name is Woodstock. My dad was conceived at the concert and he like my grandparents, have been on drugs all their life. My dad is called Woody. I am the odd one out in our family. I like fresh air good food and lots of exercise. It is hard growing up when your parents don't understand you. I just get on my bike and ride with Mr. Patton whenever I can. Mr. Patton got me a really good deal on the bike. It was much better than what we ordered. In the winter I cross country ski on all the snowmobile trails in the area. I have two part time jobs. I deliver the Peterborough Examiner after school Monday to Saturday and my second job is in Ferguson's garage. He fixes cars, trucks, and small engines. I clean up after hours. I clean and put away all the tools, sweep the floor, clean the washroom. I soon will be giving up the paper route as Mr. Ferguson wants me to start working at his horse farm just outside of Norwood. He pays me the minimum wage plus each month I get a big bonus if he has been busy. I flatten all the boxes and sort all the garbage into the recycle bins. The shop has a great sound system so I play all my favourites. Sometimes when my mom and dad run out of money they borrow mine. They say they will pay it back but they never do. I have a bank account at the Royal and I will have enough to go to college or university. I want to be in business like you and travel and meet interesting and different people. You like what you do and it sure shows that you are having fun talking to us. I am very good in math and science. I would like to work in a business helping our Planet. That would be a wish to big. Mr. Ferguson says I can bring my younger brother to the garage. He will pay him. I won't tell mom and dad Mervin is getting paid and I will open up an account for him. The teller at the bank is supposed to have our parents co-sign. She just put an X on the line and put her initial beside my account.

Trippage is easy for me. A car comes in for an oil change covered in snow and ice and Mortimer is under the back wheel in ice and brown salt. As the heat of the garage warms up the car the water starts to drop. Mechanics need to wear rain coats in the winter. Mortimer falls on Mr. Ferguson's overalls. Later that day the overalls are hung on the hook next to the heater. Mortimer is vaporized and is breathed in by Sparky the watch dog. When Sparky is let out the next day Mortimer finds himself on the side of a tree on King Street. The sun comes out and Mortimer is vaporized and off again. He catches the kite being flown by Bob Rae and does he get dizzy loop to looping around. Again he gets vaporized. Mortimer attaches to a dust particle near Pendleton and falls into the garden of Gus and Olive Brackenridge. Gus grows the biggest squash in the area and Mortimer gets sucked up into the giant squash/pumpkin. In the fall they carefully put wide straps under the squash. The fork lift loads the squash onto a special pallet and then onto the flat bed truck for the contest. This year Gus came in second but all the squash seeds are saved and Mortimer is trapped in one of those seeds. His new owners are from Elliot Lake and they are going to be planting the seeds in their window box next March.

I have to do my other homework.

Could you teach us how to draw on the black board when you come back?

Yours truly,

Woodstock Sheppard

Mrs. Ferguson helped me with this letter.

(Mrs. Ferguson taught Woodstock and Mervin in grade one and Mr. Ferguson's mother is the fill in teller at the Royal Bank.)

Hi Mr. Jack

Mortimer fell as a snow flake in January in our large 20 acre field next to the Hydro lines. March has arrived and the sun is out and it is still −15. The snow is disappearing. Sublimation or evaportisation is eating it down to the corn stubble. Mortimer left and is part of the clouds above. He fell as snow in Tatlock next to Clayton. Two days later the sun did it again. Sublimation takes most of the exposed snow. If he had landed in the woods he could have melted and seeped into the ground water. Mortimer fell as rain at Mount Tremblant. He landed on a ski jacket and it was put in a suitcase and is travelling back to Toronto on Porter airlines. Dad unpacked his suitcase and Mortimer evaporated and was sucked into the cold air return. He was warmed and ended up in the living room watching the television. Mortimer was sucked into the fireplace and he is out and up the chimney for another adventure.

I stuck a yardstick in our 20 acre field and we have lost over half the snow and the temperature has not been above freezing. Sublimation is a real thief and it is robbing us of ground water. You are right, we need forests to protect the snow and build ground water. Farmers need ground water.

Mr. Wilson how can we stop the sun stealing our water?

Murray Pethrick.

Ps. My dad is a millwright, and he is also a diesel mechanic and a certified welder. Grand Dad used to rebuild the car motors then diesel truck engines that gave the power to rope tows, then chair lifts. They saw it coming. That is the government. Dad was one of the first to certify for safety inspector. He is under contract with many ski centres across Canada and the States. He installs, services and inspects chair lifts and gondolas. Dad works with all the major manufacturers Dopplelmayr, Bmf and many others. He is what they call a trouble shooter. When they can't fix a problem or when there is an accident they call Dad.

Mom and I drive Dad to Peterborough airport and he flies out of there. On school holidays we get to go wherever dad is working. He sometimes gets to take a run or two with us. The last few summers he has been busy installing sip lines. Some sips are over two kilometers long through and over the tree tops. Mortimer will have fun sipping this summer. Dad has two contracts he is working on. One is in Barry's Bay, Ontario for the Jaworski Lumber Company. I sure hope it is not in black fly season.

Dad was able to sell an old chairlift system to one customer and he has designed a hanger for mountain bikes. Only one person and one bike per chair.

The same evening Jack was reading the letters sent to him, Eddie was at home reading the parent letter put on his desk at school.

⌒ᕲ

Letter to Mr. Eddie Patton

Monday morning Eddie found a letter placed on the centre of his desk. He always made a practice of waiting until the opening exercises were completed. He would carefully pick up any parent's envelope or correspondence with the full attention of the class. "Oh, I see I have another letter. I'll put it in my briefcase and read it after school when I have time to concentrate. We have a lot to do so let's get busy." He did not want to open any letter and have the messenger go home with their impression of Mr. Patton.

That evening after coming home from his bike ride, Eddie remembered the letter and went to his bag. He decided to read it before he started his marking and lesson preps for the next day.

Eddie always expected the worst but rarely did it happen or turn out that way. Eddie, like most teachers, was watching the change of parent involvement in the school system. More and more parents were looking for someone else to blame when their little Johnny or Susie turned out not to be the rocket scientist they thought they were going to be. After all they had bought the little person every toy and educational gadget they could afford and more. Eddie thought, "If they just could spend some time, not money, with their own children."

Dear Mr. Patton:

We were both surprised to see the assignment that you had given our daughter Lucy. Jim and I spent time between our two working schedules and have an estimated figure for each common drug we dispense.

Any village, town, or city that is on a sewer system has the same problem as a ship at sea.

I am researching to find a study that was conducted by Trent University students a few years ago on the Trent Severn waterway system. The students measured the water quality from the top of the system to the bottom where the water enters the Bay of Quinte, Lake Ontario. Concentration levels of all the drugs that were checked increased as they went downstream. It's in my files somewhere. When I find the study I will make a copy for you and send it in with Lucy.

This is the first time we have seen our two daughters the slightest bit interested in the pharmaceutical business. This topic has really caught their attention and rightly so.

Jim is our representative on our local organization of pharmacists. Jim is planning on bringing your assignment to the attention of the conference planning committee. He will also provide them with the quantities of drugs that we think are entering the water stream in the two small towns we work in. This is only the tip of the iceberg, Mr. Patton. We are part of the distribution channel and part of the problem.

Drug manufacturers research and, or buy drugs they can patent. Monopoly profits provide funding for a growing drug industry. Doctors and specialists are hustled to recommend their products. Drug companies are concerned about the bottom line and don't care about where their drugs end up. They don't pay for the clean up; they just scoop up the profits at the front end and let everyone else pay the price at the end.

Doctors prescribe drugs as patients think that is the only way they can get better. They are looking for the magic pill. We fill the prescription and then they go home. The drug washes through the body into the water treatment plant and out the pipe into the water system. It all goes downhill, back to the oceans of the world.

We all watch as urine samples are taken and used to disqualify athletes. All our pharmaceutical drug users urinate and it goes eventually into our water system. We are also aware of the radionuclide pill for thyroid problems, injection or liquid radium given to patients before many types of scans. Doctors are prescribing hormonal pills, birth control pills and many other drugs. These drugs pass through the body into our water system.

I am getting to sound like my daughter. You have taught her well. She talks nonstop about your class. Jim has had to replace the tires on her bike three times this year so far.

Mr. Patton, our industry has a problem. We are part of the complete system and we are filling our waters with drugs.

Just about everyone in town is on some type of drug. We are all part of the problem.

My sister owns and operates an on-line travel agency out of her home. When she got married and moved to Tory Hill, there were no employment opportunities so she contacted her former boss in Barrie to set up a sub-agency. When we met last weekend and my two daughters got finished talking she vowed not to sell one more cruise ticket. She is going to spread the word to her clients and look for alternatives, plus she plans to contact the main office in Barrie. Wilma is like a dog with a bone. She is the start of the movement.

Mr. Patton, do you have Mr. Wilson's contact information? Wilma would like him to speak to their travel association and Jim would like him to speak to the pharmaceutical association. Jim and I would like to hear one of Mr. Wilson's presentations. Do you know when he is speaking next?

Yours truly,

Jodie and Jim Logan

P.S. Monique has convinced us both. We are turning our back five-acre field into a tree seedling growing operation.

CHAPTER EIGHT

Dougiee's Letter to Jack Wilson

Jack picked up the last student letter. He knew it would be on the bottom and saved for last. It was Dougiee's. He was anticipating something different and he was not to be disappointed. A note was paper clipped to the letter.

> After you read this you will
> know why it is nice for a teacher
> to have a Dougiee every once
> in a while.
> Barbara

Mr. Jack:

Hi, this is Dougiee.

I have done some homework about trippage and talked to Mr. Bob. Water is H2O Mort is H2, Hydrogen as he is a bit taller. Molly is O, oxygen as she is getting on in age and putting on a bit. Mort can explode and with Molly they can be fiery when they get together. Joined together they make a wonderful couple and they are enjoyed by everyone.

Mort: "Molly and I have traced back our genealogy chart and our families go back over 4 billion years. We were all together on the Mother Planet and doing very well. A few million years ago Mother Planet blew apart when it was hit by another planet. All our records were scattered at that time and it will take some work and effort to find them. The only thing left of Mother Planet is the centre and we call it the sun. Our solar system is all the pieces that once surrounded the molten core of Mother and the planet that Mother collided with.

Planet Earth just happened to have the correct amount of hot core, water and chunks of material to support some of the life that was on Mother Planet. Planet Earth got lots of water and lots of life but it was difficult and dangerous at first. I have been separated from many Mollys over the millions of years and have travelled just about everywhere more times than you would want to know. I have helped erode the rocks at Niagara Falls, dug the river bed in the Grand Canyon, carried sediment into the Mississippi River delta, and flowed down the glacier at Whistler. Mr. Jack not once but millions of times. I am very old but as healthy as the day I was created. I am getting tired of wheezing my way through China.

Most of the dinosaurs didn't survive the blast but their bodies were preserved in tar pits and oil deposits. The other life forms couldn't get at the bodies to eat them. The dinosaurs that were exposed were eaten millions of years ago. Those dead dinosaur bodies date Earth. That was the blast time that destroyed Mother and created the start of the solar system. Our Mother planet was 4 billion years old. People on earth think that is when Earth was created. Big error says Mr. Bob.

Many of my friends that went to places like the moon and other planets haven't had as much fun as the H2 and Os on Earth."

Molly: "Mort and I have been together for over 20 million years and we are still travelling together. The odd time we get stuck for a hundred or so years in a tree or a few thousand in ground water caves but those are short stops. As water vapour we have been breathed in and out by millions of people of all race and colour. Inside people look the same but they sure make a fuss about the outside. We came in a week ago from Red Deer, Alberta. We were in a piece of lumber and the kiln drove us out and away. The rain storm last week landed us on the roof of Mr. Bob's chicken house, where two pet hens, Flow and Fluff work and play. Down into the rain barrel and Mr. Bob opened the valve on the hose and we were happy to soak into the ground. The tomato roots were competing with the squash roots and we will soon be in a tomato pasta dish and gone again. Mr. Jack, we never get tired of travelling. You could call us Molly and Mort Trippage."

Yours

Dougiee

Ps. When will you be swinging by this way? At night I am practicing all your diagrams. Mr. Bob picked up a huge roll of newspaper at the Peterborough Examiner office in Peterborough. The ends of rolls are used by people packing. They stretch out across the living and dining room floor and I pretend I am at the lecture hall. Mr. Bob last week picked up three more rolls. I hope to be as fast and as good as you are some day.

Jack read Barbara's note again.

> After you read this you will
> know why it is nice for a teacher
> to have a Dougiee every once
> in a while.
> Barbara

Jack looked at Joy, "Barbara and Julie were right. You sure get to know your students one at a time."

Drop a line and let me know how you are doing.

Rod

rod@sawmillbooks.ca

www.sawmillbooks.ca

Epilogue

An interview with Rod by Roxie from CGNR.

Q. Do you prefer to be called Rodney or Rod?

A. When someone calls me Rodney I know they are from Norwood or a relative. I like to be called Rod. I say to people, it's Rod, like fishing rod. I get called Bob, Job, and Knob, anything that rhymes or sounds like Rod.

Q. You could write about many topics and use many characters, why did you choose to write about the environment? After all, your background is not in natural sciences.

A. I guess I am a bit like the main character, Jack Wilson. When I retired I bought a Woodmizer saw mill and a diesel Kubota skidder/tractor. This was a long-time wish, and now it was real. I read the manuals and started to cut trees and produce lumber. The first logs I sawed were sugar maple as the property we owned and still own was an old sugar bush. Grandfather Ed Irwin supplemented the family income with the first cash crop of the season. Maple syrup money bought the seeds for the next grain crop. I first went through the woodlot and took out all the hanging "widow-makers" and applied to the Ontario Woodlot Association for the managed forest program. The big old maple trees were starting to split and fall apart. They needed to be taken out of the bush before someone was crushed.

Green lumber coming off the mill was extremely heavy, but I muscled up and was soon piling and stickering it with no after aches and pains the next day. The next fall I had an opportunity to sell a few hundred board feet. That same green heavy lumber was now much easier to lift and move. We call that hand bombing. We filled the contractor's truck and trailer whereas with green lumber it would have taken two trips.

I knew from experience with cutting firewood that green blocks split easily but are very heavy. I noticed that the heat of the sun and the moving air would cut their weight in half. Dry seasoned firewood burns easily with far less creosote to clog up a chimney. At that time I asked myself where did the water go? This is what started my research for *Cooling Down Planet Earth*.

Q. All the articles today seem to list three reasons for the oceans filling up. You have added four more! How did you discover the other sources of water that are filling up our oceans and seas around the world?

A. At first I calculated how much water would be in a log, and then a full tree with all the branches and leaves, then this big plant had a root system that contained even more water. Internet sites are great tools for looking up statistics. The formula for calculating the surface area of a sphere is available and the percentage of the planet that is land, and the percentage of the land that was covered in trees are all available to anyone who wants to do the math. There are thousands of people around the world who know that green lumber and green wood are full of water. The trees that were here before the first cut were very large and very tall and they would make our present day forests look like shrubs. The old saw mills could handle logs six feet, two meters, in diameter. We can calculate very closely how much water we had in these storage vessels. The next calculation is quite simple. Take all that water that we lost and spread it out over the surface of the oceans and you come up with at least two inches. Then I asked myself, where did the rest of the water come from? Water is not created or destroyed. It just changes temperature and comes in different shapes. When you are working out the actual numbers your assumptions before the calculations will get you into trouble. I will leave the assumptions up to the experts.

Q. In your book you use the term first cut, not the original forests. Why?

A. There is no such thing as an original forest. Trees, plants, animals all grow and die and hopefully regenerate. When Europeans came to North America they had the tools to cut down the forests. By the time Europeans or Vikings came to this continent they had stripped most of their trees off their land and were farming. That was the first cut. Natives or first generation, I use the term first immigrants, as our archeological studies show, they arrived here from Europe or Asia a few thousand years before the next wave of Europeans. They learned to burn the forests to regenerate growth. In a mature forest, hunters and gatherers would starve to death if they did not move constantly to find enough food to survive. Burning the forests allowed the first people to settle for awhile until they over ate their location. Our early history in Ontario is that the Crown owned all the large pine trees as they were needed for sailing masts. Taller masts, more sail, made for faster ships and large Canadian white pines were the property of the British Crown.

Going back to your question; how did I discover the other sources of water that were filling up the oceans?

Green Vegetation Water Sinks, GVWS can account for an ocean rise of at least two inches. I asked myself, if polar ice caps and glaciers are melting and receding today while oceans and seas have been rising for a long time, then where did the water come from? Well, my grandfather owned a farm just east of Norwood, Ontario. A large esker ran through his two hundred acres. This is one long pile of gravel from the last ice age. The gravel at first was used on the farm then a pit was started and the process of removing all the gravel began. The farm had two small ponds and a very good dug well. The two ponds soon dried up and the well eventually went dry. It wasn't hard to connect the dots on that one. The field south of Number Seven highway was very wet and my grandfather Andrews and my dad dug by hand a trench to the Ouse River, put in red clay tile, and drained the water. My uncle later on had a mink ranch and a pig farm on that land. You can appreciate every spring with melting snow and rain what drained into the Ouse River then downstream into Rice Lake.

This was a memory of growing up and sure enough when you look at concrete and how it is made you can calculate how much water has been squeezed out of aggregate. Romans discovered and made concrete using cement. Cement is made from burning limestone and my grandfather Lain on my mother's side had a cement oven, a granite lined pit on his farm next to Crow River not far from Pethericks corner, Ontario. Making asphalt cooks then squeezes out the water in aggregate. Concrete and asphalt are the two most common building materials used in the world. And don't forget clay bricks as they cook out the water to make bricks. Think of the mountains of coal around the world. They held water as well, and all of this water today is pulled back to the ocean by gravity. Water runs down hill.

Reducing the natural sponges, or aggregate of the world can account for at least two inches. It all adds up. Activities such as draining wet lands and draining farm fields have taken place and continue today around the world. Florida and Holland are two obvious examples. Drying out peat bogs for fuel dumps even more water into the oceans. Trent Severn Water has built their headquarters and storage yard on a wetland. That complete flat area on Armour Road in Peterborough was originally a wetland. It was filled in. Today I would call it flood land. Trent Severn has a problem with flooding every spring. Their Peterborough head office is sitting on wetland that historically stored water during the spring runoff. Someone is not connecting the dots.

Another source of water is the underground water system that can be found worldwide. Aquifers around the world are being depleted. The water has to end up somewhere. Eventually all fresh water is pulled back to the oceans and seas of the world. We are pumping out ground water faster than it is replenishing itself. We are not helping out the system. Spring and rainy season run off could be put back into the ground water table but we haven't caught on. We need to start putting back the water. Those sink holes around the world are signs that we are taking out too much ground water. We like to blame the sink holes on some other cause.

Q. Why have geographers, scientists and environmentalists not zeroed in on these causes, or I should say sources that increase the ocean levels?

A. A good question. I think because if it is not visual therefore it is not obvious. Out of sight out of mind is an old expression, but I think it fits in this case. I have discussed with friends about the water in trees and how we have cut them down and how this water is filling up the oceans. Their reaction is of dismay. They have started to look at trees differently.

Q. You wrote a book and told a story, why did you decide to use this approach to making your information known as opposed to publishing your findings?

A. It can be summed up in three words: Lack of credibility!

I am just going to go sideways for a moment... A couple of years ago I had a health scare. My wife, Claire and I were sitting in one of the many waiting rooms at Toronto General Hospital when I happened to pick up a magazine. It was

Scientific American. On the front cover was the story about Dr. Judith Curry. Here was a professional being called a heretic because she was talking to people outside the environmental industry and questioning what was going on. It struck me then - what chance would I have with my findings? Dr. Judith Curry, who heads the School of Earth and Atmospheric Sciences at the Georgian Institute of Technology, has been an inspiration for me to continue my research. I hope we will meet someday. I would like to thank her and maybe give her a big farm hug.

The reality is that I have no credentials in this industry. When I talk to people about sawing lumber my word is credible. I chose to write a book that would have wide appeal and be easily understood by an age group as young as grade six. Young and old alike could read and try out the experiments to prove it to themselves. Our environmental, science, and medical industries are huge and very conservative. Unless a recognized expert gives the stamp of approval ideas can sit on the back burner for years. A good example is in my first book in the chapter dealing with the car industry. The defect is still being built into just about every car and light truck in the world today. How do you move the auto giants? One auto engineer said to me after seeing the video on YouTube that it would be easier to change the mafia than the auto industry!

Q. You approached global warming through the backdoor. *Cooling Down Planet Earth* no one to my knowledge has looked at how the planet keeps itself cool. How did you think of the idea?

A. I guess like everyone else I was tired of reading the same old story over and over again. One can almost predict the contents of a magazine or newspaper article. I thought to myself why not look at how the planet cools itself. We have had a few ice ages where clearly humans were not on the hook for those events. I thought to myself why not look at how the Planet cools itself.

To make it simple and very easy for young minds to grasp, the term SEA came about - Shade, Evaporation and Altitude. There are only three ways the planet can cool itself and it became clear to me that we are standing in the way!

Q. You are not a professional author and you have a full schedule. How do you find time to write a book let alone two? Did I hear rumblings about a third book on its way?

A. Yes, I have finished my third book. By special request from my readers, the characters who were in the first book will be back. For those asking about sex kitten Ruthie, yes, she is a real person. And yes she will be back. I am presently struggling with the title as it is a mix of down home stories and environmental messages that are calling for our attention.

Q. One last question. Why did you self-publish?

A. When I started to write my first book I could not get an agent let alone a publisher. There is no shortage of books on the shelf. Once writers have established a track record, or have sold books to demonstrate that their work sells, then they are more likely to get professional help. My first book was finished and I contacted Greg Ioannou, a very well recognized editor. He asked for a sample chapter of my first book. Greg edited the Lemon Aid car guides so I sent him the short story dealing with the defect in the auto industry. A couple of weeks later he took the time to call me back. He was most gracious and gentle. He said a book of short stories is not in demand and that it would be a very tough sell in the market place. He suggested that I would be wasting my money to have him edit the short stories. Initially I was very disappointed, much like being a teenager at a sock hop and the girl you ask to dance turns you down. I got over that quickly and did a little self-assessment. I knew I had received valuable information when I found out later that Greg was taking his turn as President of the Canadian Editors' Association. I was grateful for his input. My short stories were melded into chapters and *Ten Bridges Seven Churches No Stop Light* became my first book. I am the bestselling author in Lakehurst and now working on Lakefield. My wife, Claire reminds me every once in a while that Lakehurst is a hamlet which has yet to open up a book store. Maybe someday I will be lucky enough to have a publisher. Oh, I define luck as preparation meeting opportunity, to quote Gary Adamson, a former work colleague. His definition stuck with me.

Finding a good publisher to spread the word around the world would be a dream come true. When I talk to young grade six students, I call them sixers. I show them a globe to demonstrate how our planet is just a big space ship or airplane on a long trip. There are some first class seats, some business class and some jammed in steerage accommodation. Below the seat level is a lot of baggage. We are travelling together, going to the same place at the same time, and if someone is smoking at the back or front we all suffer from second hand smoke. Using such an analogy helps students understand that burning coal in the USA, China or India affects us all. A tsunami that hits Japan and later washes up wreckage on our Canadian west coast is another example I use.

Q. What advice would you give to new budding authors?

A. I have had several budding authors call me for advice and my answer is not necessarily what they want to hear. The book stores have no shortage of books; grocery stores have no shortage of products. I explain how to set up their manuscript for self-publishing and then I drop the bomb: spend an hour a day and enjoy writing. At the end of the process if you sell four hundred copies of what you think is the best thing since sliced bread you are the top of the heap. Then to break the silence on the end of the phone, I tell them that I purchased a book on how to play hockey. I finished it and did all the drills and exercises and I didn't make it through the pee wee cut... the NHL is a few more books away.

Q. Thanks Rod for your time, and one last question. I belong to a book club. Would you come out and talk to us some evening?

A. Be glad to. You know my e-mail, just drop a line.

Q. Oops I see my note pad. I almost forgot to ask you about the Rain Drop Program.

A. I am really excited about it. What would make most happy is to get people – young and old – to visualize the concept that trees are full of water. Trees are a very precious resource. The rain drop is a vehicle to catch people's attention, get the message across, and to keep a visual reminder that we can make a difference. It's not a fundraiser. It makes it so easy for people to get involved. It all starts, one rain drop at a time.

Rod reached into a cardboard box by his desk and pulled out a piece of freshly cut white pine and a rain drop wrapped around it. 270 grams was printed across the top face of the piece of pine. "Will you take this green piece of pine home and place it somewhere in your house to dry? Weigh it after a few weeks then again each month and record the results. This is the simplest way to witness or demonstrate the amount of water in trees. Please let me know what issue the interview will be in and I will be interested to hear the reaction you get when you hang the rain drop on your chosen tree.

Rod; I hope I haven't talked too much.

Roxie from CGNR; Don't worry my editor will likely cut this interview down to one page. You will recognize the interview, but it will be short when it hits the press.

About The Rain Drop Program

This is one visible way to grab people's attention and build awareness that we all must start now to put the water back.

Each community can adopt the Rain Drop Program and every citizen can participate. The rain drop is placed on the largest water sink (tree). It is made from non-rusting material and on the back the date and volume of water can be recorded. The rain drop is designed to last a very long time so it can be moved from tree to tree as needed. Avoid using screws or nails to place or hang the rain drop. Using a ribbon or cord will prevent future problems when a tree needs to be cut down. Future sawyers and arborists will thank you for it.

When trees are taken down... **splash not timber is the call!**

To find out more about rain drops, visit www.sawmill books.ca

www.ingramcontent.com/pod-product-compliance
Lightning Source LLC
Chambersburg PA
CBHW071711170526
45165CB00005B/1966